DIRECT INTEGRAL THEORY

PURE AND APPLIED MATHEMATICS

A Program of Monographs, Textbooks, and Lecture Notes

Executive Editors

Earl J. Taft
Rutgers University
New Brunswick, New Jersey

Edwin Hewitt
University of Washington
Seattle, Washington

Chairman of the Editorial Board

S. Kobayashi
University of California, Berkeley
Berkeley, California

Editorial Board

Masanao Aoki
University of California, Los Angeles

Zuhair Nashed
University of Delaware

Glen E. Bredon
Rutgers University

Irving Reiner
University of Illinois at Urbana-Champaign

Sigurdur Helgason
Massachusetts Institute of Technology

Paul J. Sally, Jr.
University of Chicago

G. Leitmann
University of California, Berkeley

Jane Cronin Scanlon
Rutgers University

Marvin Marcus
University of California, Santa Barbara

Martin Schechter
Yeshiva University

W. S. Massey
Yale University

Julius L. Shaneson
Rutgers University

Olga Taussky Todd
California Institute of Technology

Contributions to *Lecture Notes in Pure and Applied Mathematics* are reproduced by direct photography of the author's typewritten manuscript. Potential authors are advised to submit preliminary manuscripts for review purposes. After acceptance, the author is responsible for preparing the final manuscript in camera-ready form, suitable for direct reproduction. Marcel Dekker, Inc. will furnish instructions to authors and special typing paper. Sample pages are reviewed and returned with our suggestions to assure quality control and the most attractive rendering of your manuscript. The publisher will also be happy to supervise and assist in all stages of the preparation of your camera-ready manuscript.

LECTURE NOTES
IN PURE AND APPLIED MATHEMATICS

1. *N. Jacobson*, Exceptional Lie Algebras
2. *L.-Å. Lindahl and F. Poulsen*, Thin Sets in Harmonic Analysis
3. *I. Satake*, Classification Theory of Semi-Simple Algebraic Groups
4. *F. Hirzebruch, W. D. Newmann, and S. S. Koh*, Differentiable Manifolds and Quadratic Forms
5. *I. Chavel*, Riemannian Symmetric Spaces of Rank One
6. *R. B. Burckel*, Characterization of C(X) Among Its Subalgebras
7. *B. R. McDonald, A. R. Magid, and K. C. Smith*, Ring Theory: Proceedings of the Oklahoma Conference
8. *Y.-T. Siu*, Techniques of Extension of Analytic Objects
9. *S. R. Caradus, W. E. Pfaffenberger, and B. Yood*, Calkin Algebras and Algebras of Operators on Banach Spaces
10. *E. O. Roxin, P.-T. Liu, and R. L. Sternberg*, Differential Games and Control Theory
11. *M. Orzech and C. Small*, The Brauer Group of Commutative Rings
12. *S. Thomeier*, Topology and Its Applications
13. *J. M. López and K. A. Ross*, Sidon Sets
14. *W. W. Comfort and S. Negrepontis*, Continuous Pseudometrics
15. *K. McKennon and J. M. Robertson*, Locally Convex Spaces
16. *M. Carmeli and S. Malin*, Representations of the Rotation and Lorentz Groups: An Introduction
17. *G. B. Seligman*, Rational Methods in Lie Algebras
18. *D. G. de Figueiredo*, Functional Analysis: Proceedings of the Brazilian Mathematical Society Symposium
19. *L. Cesari, R. Kannan, and J. D. Schuur*, Nonlinear Functional Analysis and Differential Equations: Proceedings of the Michigan State University Conference
20. *J. J. Schäffer*, Geometry of Spheres in Normed Spaces
21. *K. Yano and M. Kon*, Anti-Invariant Submanifolds
22. *W. V. Vasconcelos*, The Rings of Dimension Two
23. *R. E. Chandler*, Hausdorff Compactifications
24. *S. P. Franklin and B. V. S. Thomas*, Topology: Proceedings of the Memphis State University Conference
25. *S. K. Jain*, Ring Theory: Proceedings of the Ohio University Conference
26. *B. R. McDonald and R. A. Morris*, Ring Theory II: Proceedings of the Second Oklahoma Conference
27. *R. B. Mura and A. Rhemtulla*, Orderable Groups
28. *J. R. Graef*, Stability of Dynamical Systems: Theory and Applications
29. *H.-C. Wang*, Homogeneous Banach Algebras
30. *E. O. Roxin, P.-T. Liu, and R. L. Sternberg*, Differential Games and Control Theory II
31. *R. D. Porter*, Introduction to Fibre Bundles
32. *M. Altman*, Contractors and Contractor Directions Theory and Applications
33. *J. S. Golan*, Decomposition and Dimension in Module Categories
34. *G. Fairweather*, Finite Element Galerkin Methods for Differential Equations
35. *J. D. Sally*, Numbers of Generators of Ideals in Local Rings
36. *S. S. Miller*, Complex Analysis: Proceedings of the S.U.N.Y. Brockport Conference
37. *R. Gordon*, Representation Theory of Algebras: Proceedings of the Philadelphia Conference
38. *M. Goto and F. D. Grosshans*, Semisimple Lie Algebras
39. *A. I. Arruda, N. C. A. da Costa, and R. Chuaqui*, Mathematical Logic: Proceedings of the First Brazilian Conference
40. *F. Van Oystaeyen*, Ring Theory: Proceedings of the 1977 Antwerp Conference

41. *F. Van Oystaeyen and A. Verschoren,* Reflectors and Localization: Application to Sheaf Theory
42. *M. Satyanarayana,* Positively Ordered Semigroups
43. *D. L. Russell,* Mathematics of Finite-Dimensional Control Systems
44. *P.-T. Liu and E. Roxin,* Differential Games and Control Theory III: Proceedings of the Third Kingston Conference, Part A
45. *A. Geràmita and J. Seberry,* Orthogonal Designs: Quadratic Forms and Hadamard Matrices
46. *J. Cigler, V. Losert, and P. Michor,* Banach Modules and Functors on Categories of Banach Spaces
47. *P.-T. Liu and J. G. Sutinen,* Control Theory in Mathematical Economics: Proceedings of the Third Kingston Conference, Part B
48. *C. Byrnes,* Partial Differential Equations and Geometry
49. *G. Klambauer,* Problems and Propositions in Analysis
50. *J. Knopfmacher,* Analytic Arithmetic of Algebraic Function Fields
51. *F. Van Oystaeyen,* Ring Theory: Proceedings of the 1978 Antwerp Conference
52. *B. Kedem,* Binary Time Series
53. *J. Barros-Neto and R. A. Artino,* Hypoelliptic Boundary-Value Problems
54. *R. L. Sternberg, A. J. Kalinowski, and J. S. Papadakis,* Nonlinear Partial Differential Equations in Engineering and Applied Science
55. *B. R. McDonald,* Ring Theory and Algebra III: Proceedings of The Third Oklahoma Conference
56. *J. S. Golan,* Structure Sheaves over a Noncommutative Ring
57. *T. V. Narayana, J. G. Williams, and R. M. Mathsen,* Combinatorics, Representation Theory and Statistical Methods in Groups: YOUNG DAY Proceedings
58. *T. A. Burton,* Modeling and Differential Equations in Biology
59. *K. H. Kim and F. W. Roush,* Introduction to Mathematical Consensus Theory
60. *J. Banas and K. Goebel,* Measures of Noncompactness in Banach Spaces
61. *O. A. Nielsen,* Direct Integral Theory

Other Volumes in Preparation

DIRECT INTEGRAL THEORY

Ole A. Nielsen

Department of Mathematics and Statistics
Queen's University
Kingston, Ontario, Canada

MARCEL DEKKER, INC. New York and Basel

Library of Congress Cataloging in Publication Data

Nielsen, Ole A.
  Direct integral theory.

  (Lecture notes in pure and applied mathematics ; 61)
  Bibliography: p.
  Includes indexes.
  1. Von Neumann algebras. 2. Representations of algebras. 3. Integrals. I. Title.
  QA326.N53   512'.55   80-21962
  ISBN 0-8247-6971-6

COPYRIGHT © 1980 by MARCEL DEKKER, INC.   ALL RIGHTS RESERVED

Neither this book nor any part may be reproduced or transmitted in any form or by any means, electronic or mechanical, including photocopying, microfilming, and recording, or by any information storage and retrieval system, without permission in writing from the publisher.

MARCEL DEKKER, INC.

270 Madison Avenue, New York, New York 10016

Current printing (last digit):

10 9 8 7 6 5 4 3 2 1

PRINTED IN THE UNITED STATES OF AMERICA

PREFACE

The summands of a direct sum are usually simpler objects than the direct sum itself, and therefore one can often learn something about some object by decomposing it into a direct sum and then studying the summands. This is especially true in finite-dimensional situations where there is a natural limit to the number of possible summands. In studying von Neumann algebras or representations of either involutive Banach algebras or locally compact groups acting on a not necessarily finite-dimensional Hilbert space, however, direct sum decompositions are of only limited use. What are much more useful are "continuous direct sums," i.e., direct integrals. In fact, associated with every direct sum or direct integral decomposition of a von Neumann algebra or of such a representation is a certain abelian algebra (the so-called algebra of diagonalizable operators) and the decomposition will be a direct sum precisely if every projection in this algebra is a sum of minimal projections in this algebra. This latter condition is always fulfilled if the underlying Hilbert space is finite-dimensional, but, in general, is not fulfilled.

The objects to which direct integral theory is applicable are von Neumann algebras acting on separable Hilbert spaces on the one hand and representations on separable Hilbert spaces of either separable involutive Banach algebras or separable locally compact groups on the other. In either case, one attempts to either decompose a given object into simpler objects or else to build up a complicated object from simpler objects. But what are these "simpler" objects? For von Neumann algebras it is the factors (i.e., those with one-dimensional centers) which cannot be decomposed as direct sums, and for representations it is the irreducible ones which cannot be decomposed as direct sums. So a "simple" von Neumann algebra ought to be a factor and a "simple" representation ought to be an irreducible one. This turns out to be true for von Neumann algebras but not quite true for representations. The reason for this state of affairs concerning representations is that while one can always decompose a given representation into a direct integral of irreducible ones, one cannot, in general, do so uniquely. One can, however, always decompose a given representation into an essentially unique direct integral of factor representations. So one tends to regard a "simple" representation as being a factor representation. Actually, for many involutive Banach algebras and locally compact groups every factor representation is a multiple of an irreducible one, so for these algebras and groups the "simple" representations are the irreducible ones after all.

The theory of Borel spaces, which is the subject of Chapter 1, is an indispensible tool in the development of direct integral theory. The task of writing an account of the theory of Borel spaces was considerably shortened by not giving complete proof of the deeper results of the theory, but, instead, of reducing them to theorems in other textbooks. Chapters 2, 3, and 4 are similar to one another in that they are concerned with the existence, uniqueness, elementary properties, and examples of direct integral decompositions of Hilbert spaces and operators, of representations of involutive Banach algebras and locally compact groups, and of von Neumann algebras, respectively. Chapter 5 is devoted to the relationship between direct integral decompositions and types of von Neumann algebras, and Chapter 6 to a detailed analysis of representations by means of direct integral decompositions. Finally, the three appendices are intended to acquaint the reader with as much of the theory of von Neumann algebras and of representations of involutive Banach algebras and locally compact groups as is needed for reading this book.

The contents of Sections 1-15, 18-20, and 26-28 are for the most part well-known, and the expert will not find anything new here. A feature of these sections which is worth pointing out is that the direct integral theory of von Neumann algebras is made to depend logically on that of representations. On the other hand, the contents of Sections 17 and 21-25 are not so well-known, consisting, as it does, of Effros' approach to the direct integral of von Neumann algebras and of the author's proofs of the theorems relating direct integrals and types of von Neumann algebras. It should be called to the attention of the experts that while these proofs involve modular automorphism groups they do not require (as did earlier proofs) infinite traces, fields of traces, or fields of Hilbert algebras.

PREFACE

This book is self-contained except for the proofs of some of the results stated in Chapter 1 and Appendix A. The reader is assumed to have some knowledge of measure theory and of the theory of Hilbert spaces and bounded operators thereon (up to and including the spectral theorem for self-adjoint operators), and, for Chapters 4 and 5, an acquaintance with the theory of von Neumann algebras.

During the 1969-1970 academic year the author was privileged to be able to attend the lectures on the direct integral theory of separable involutive Banach algebras which Professor J. M. G. Fell gave at the University of Pennsylvania. The influence of these lectures is especially strong in Sections 2, 4, 27, and 28.

The genesis of these lecture notes was a course taught by the author at Aarhus University during the spring of 1975. These lectures covered the present Chapters 1-3 and Appendices B and C. The author would like to thank the Mathematics Institute of Aarhus University for giving him an opportunity to lecture on direct integral theory; the Canadian NRC and NSERC, and the Danish SNF for their financial support; Dita Andersen and Ulla Jacobsen for typing the first draft of the manuscript; and Marge Lambert for retyping the manuscript according to the publisher's specifications.

O.A.N.

CONTENTS

Preface   iii

Chapter 1
BOREL SPACES   1

1. Definitions   1
2. Borel Spaces and Functions   3
3. Borel Spaces and Equivalence Relations   7
4. Borel Spaces and Measures   9
   Historical Comments   12

Chapter 2
DIRECT INTEGRALS OF HILBERT SPACES AND OPERATORS   15

5. Hilbert Space-valued L^2-spaces   15
6. Operators on Hilbert Space-valued L^2-spaces   17
7. Fields of Hilbert Spaces and Operators   22
8. The Construction of Coherences   27
9. Direct Integral Decompositions   30
10. Examples   34
    Historical Comments   42

Chapter 3
DIRECT INTEGRALS OF REPRESENTATIONS   45

11. Definitions and Some Elementary Properties   45
12. Equivalence, Existence and Uniqueness of Direct
    Integral Decompositions   47
13. Maximal and Central Decompositions   51
14. Some Applications   54
15. Examples   55
    Historical Comments   62

Chapter 4
DIRECT INTEGRALS OF VON NEUMANN ALGEBRAS — 63

16. Hausdorff Metrics — 63
17. The Effros Borel Structure — 67
18. Definitions and Some Elementary Properties — 74
19. Some Further Properties — 76
20. Examples — 80
    Historical Comments — 82

Chapter 5
DIRECT INTEGRALS AND TYPES OF VON NEUMANN ALGEBRAS — 85

21. The Statements of Two Theorems — 85
22. Partial Proof of Theorem 21.1 — 87
23. Proof of Theorem 21.2 — 93
24. Some Technical Lemmas — 99
25. The Completion of the Proof of Theorem 21.1 — 112
    Historical Comments — 113

Chapter 6
MEASURES AND REPRESENTATIONS — 115

26. The Dual and the Quasi-daul — 115
27. Measures on the Dual and Representations — 120
28. Measures on the Quasi-dual and Representations — 126
    Historical Comments — 135

Appendix A
VON NEUMANN ALGEBRAS — 139

Appendix B
REPRESENTATIONS OF INVOLUTIVE BANACH ALGEBRAS — 145

Appendix C
REPRESENTATIONS OF LOCALLY COMPACT GROUPS — 151

REFERENCES — 157
INDEX OF NOTATION — 161
INDEX — 163

DIRECT
INTEGRAL
THEORY

Chapter 1
BOREL SPACES

1. Definitions

A *Borel structure* on a set X is a σ-field of subsets of X, i.e., a family of subsets of X containing X itself and closed under complements and countable unions. A *Borel space* is a pair (X,S) consisting of a set X and a Borel structure S on X; when no confusion can arise one writes X in place of (X,S) and calls the members of S the *Borel subsets* of X. A Borel function from a Borel space X into a Borel space Y is a function f from X to Y with the property that $f^{-1}(A)$ is a Borel set in X whenever A is a Borel set in Y. Two Borel spaces are called *Borel isomorphic* if there is a *Borel isomorphism* of one onto the other, i.e., if there is a one-one Borel function of one onto the other whose inverse is also a Borel function.

It is trivial that the intersection of any number of Borel structures on a set X is again a Borel structure on X. Thus each topological space can (and will) be regarded as a Borel space by taking the smallest Borel structure containing the topology. If X is a set, if $(X_i)_{i \in I}$ is a family of Borel spaces, and if, for each $i \in I$, f_i is a function from X to X_i [respectively, X_i to X] then the Borel structure on X *generated by the* f_i *and the* X_i, $i \in I$, is the smallest [respectively, largest] one making each f_i Borel. The *relative Borel structure* on a subset Y of a Borel space X is that generated by i and X, where i is the inclusion mapping of Y into X, and the *Borel subspace of* X *based on* Y is Y together with this Borel structure. If $(X_i)_{i \in I}$ is a family of [respectively, a family of mutually disjoint] Borel spaces then the *product Borel structure* on $\prod_{i \in I} X_i$ [respectively, *sum Borel structure* on $\cup_{i \in I} X_i$] is that generated by the p_i [respectively, the q_i] and the

X_i, $i \in I$, where p_i is the projection map of $\Pi_{i \in I} X_i$ onto X_i [respectively, q_i is the inclusion map of X_i into $\cup_{i \in I} X_i$]; $\Pi_{i \in I} X_i$ [respectively, $\cup_{i \in I} X_i$] with this Borel structure is called the *product Borel space* [respectively, *sum Borel space*] of the X_i. If $\sim$ is an equivalence relation on a Borel space X then the *quotient Borel structure* on the set X/$\sim$ of equivalence classes is that generated by ω and X, where ω is the quotient mapping of X onto X/$\sim$, and the *quotient Borel space of X by* $\sim$ is X/$\sim$ together with this Borel structure.

Suppose that $\sim$ is an equivalence relation on a set X, and consider the set X/$\sim$ of equivalence classes and the quotient mapping ω of X onto X/$\sim$. A *transversal* for $\sim$ is a subset A of X such that the restriction of ω to A is one-one and onto X/$\sim$, and a *cross-section* for $\sim$ is a map σ from X/$\sim$ to X such that $\omega \circ \sigma$ is the identity on X/$\sim$. There is, of course, a one-one correspondence between transversals and cross-sections; however, if X is a Borel space and if X/$\sim$ is given the quotient Borel structure, then there is not in general a one-one correspondence between Borel transversals and Borel cross-sections (cf. Proposition 3.4).

A Borel space is called *countably separated* if there are countably many Borel sets in the space which separate the points of the space, and *countably generated* if there are countably many Borel sets in the space which both separate the points of the space and generate the Borel structure.

A topological space is called *Polish* if it is separable and if its topology is generated by a complete metric. A Borel space X is called *standard* if there is a Polish space Z and a Borel subset Y of Z such that X is Borel isomorphic to the Borel subspace of Z based on Y. A Borel space is called *analytic* if it is countably generated and is the image of a standard Borel space by a Borel function. Finally, a subset of a Borel space is called *standard* or *analytic* if the subspace based on it is standard or analytic, respectively, and *coanalytic* if its complement is analytic.

A Borel structure S on a set X will be called *countably separated* [respectively, *countably generated, standard, analytic*] if the corresponding Borel space (X,S) is countably separated [respectively, countably generated, standard, analytic].

A *Borel measure* on a Borel space X is a countably additive σ-finite measure defined on the Borel sets of X and taking values in $[0,\infty]$. A subset A of a Borel space X is called μ-*measurable*, where μ is some Borel measure on X, if there are Borel sets B and C in X and $B \subset A \subset C$ and $\mu(C - B) = 0$, and *universally measurable* if it is μ-measurable for every Borel measure μ on X. A Borel measure μ on a Borel space X is called *standard* if there is a standard Borel set A in X with $\mu(X - A) = 0$.

2. Borel Spaces and Functions

The first three propositions do not really belong to the theory of Borel spaces, but are nevertheless included here since they will be useful and since this seems to be the most appropriate place to state them.

PROPOSITION 2.1. Every second countable locally compact space is Polish. A subset of a Polish space is Polish in its relative topology if and only if it is a G_δ.

Proof. This is well-known (see [3, p. 43, Corollaire and p. 122, Corollaire] for the first assertion and [3, p. 123, Théorème 1] for the second, for example). □

PROPOSITION 2.2 (Monotone Class Theorem). If a monotone class of subsets of a set X (i.e., a collection of subsets of X closed under unions and intersections of monotone sequences) contains a field of subsets of X (i.e., a collection of subsets of X containing X itself and closed under complements and finite unions) then it also contains the Borel structure generated by that field.

Proof. This too is well-known (see [25, §6, Theorem B], for example). □

PROPOSITION 2.3. If X is a separable metric space, if Y is a set, if Z is a metric space, and if f is a function from X × Y to Z then

  (a) if Y is a topological space and if f is separately continuous then $f^{-1}(F)$ is a G_δ in X × Y for each closed set F in Z, and

  (b) if Y is a Borel space, if $f(\cdot,\xi)$ is continuous for each fixed ξ in Y, and if $f(\zeta,\cdot)$ is Borel for each fixed ζ in X, then f is Borel.

Proof. It is enough to observe that if (ζ_n) is a dense sequence in X and if d_1 and d_2 are metrics on X and Z, respectively, defining the given topologies then

$$f^{-1}(F) = \bigcap_{k=1}^{\infty} \bigcup_{n=1}^{\infty} \{(\zeta,\xi) \in X \times Y: d_1(\zeta,\zeta_n) < k^{-1} \text{ and } d_2(f(\zeta_n,\xi),F) < k^{-1}\}$$

for any closed subset F of Z. □

The next lemma turns out to be indispensable in developing the theory of Borel spaces.

LEMMA 2.4. A Borel structure is countably generated if and only if it is generated by a separable metric topology. If (X_i) is a countable family of separable metric spaces then the product Borel structure on $\Pi_i X_i$ is generated by the product topology. If X is a countably generated Borel space then, with respect to the product Borel structure, the diagonal D in the product of X with itself at most countably many times is a Borel set and the natural mapping of X onto D is a Borel isomorphism of X onto the Borel subspace based on D.

Proof. The proof of the second assertion and half of the first are trivial, and the third follows easily from the first two. Suppose that X is a countably generated Borel space, and let (S_n) be a sequence of Borel subsets of X which separates the points of X and generates the Borel structure. Then the map f from X to $\{0,1\}^{\mathbb{N}}$ defined by putting

$$f(\zeta)_n = \begin{cases} 0 & \text{if } \zeta \in S_n \\ 1 & \text{otherwise} \end{cases}$$

for each ζ in X and each $n \in \mathbb{N}$ is a Borel isomorphism of X onto the Borel subspace of $\{0,1\}^{\mathbb{N}}$ based on $f(X)$. □

LEMMA 2.5. The product [respectively, sum] Borel space of a countable family (X_i) of Borel spaces [respectively, of mutually disjoint Borel spaces] is standard if each X_i is standard and analytic if each X_i is analytic.

Proof. This follows easily with the aid of the preceeding lemma. □

PROPOSITION 2.6. If (S_n) is a sequence of subsets of a countably generated Borel space then $\cup_n S_n$ is standard if each S_n is standard and Borel, $\cap_n S_n$ is standard if each S_n is standard, and both $\cup_n S_n$ and $\cap_n S_n$ are analytic if each S_n is analytic.

Proof. The first assertion follows from Lemma 2.5 since, in this case, one may assume that the S_n are mutually disjoint. If the S_n are all analytic then for each n there is a standard Borel space Y_n and a Borel mapping f_n of Y_n onto S_n. Then

$$T = \{(\xi_n) \in \Pi_{n=1}^{\infty} Y_n : f_1(\xi_1) = f_2(\xi_2) = \cdots \}$$

2. BOREL SPACES AND FUNCTIONS

is a Borel subset of $\prod_{n=1}^{\infty} Y_n$ by Lemma 2.4 and $(\xi_n) \mapsto f_1(\xi_1)$ is a Borel mapping, say g, of T onto $\cap_{n=1}^{\infty} S_n$, and hence $\cap_{n=1}^{\infty} S_n$ is analytic. Now the Y_n can, of course, be taken to be mutually disjoint. Then the mapping of $\cup_{n=1}^{\infty} Y_n$ onto $\cup_{n=1}^{\infty} S_n$ whose restriction to Y_n is f_n for each n is Borel with respect to the sum Borel structure on $\cup_{n=1}^{\infty} Y_n$, and so $\cup_{n=1}^{\infty} S_n$ is analytic by Lemma 2.5. Finally, if the S_n are all standard one can take Y_n to be S_n and f_n to be identity, and then g is a Borel isomorphism of T onto $\cap_{n=1}^{\infty} S_n$ by Lemma 2.4. □

PROPOSITION 2.7. If f is a Borel mapping of an analytic Borel space X into a countably generated Borel space Y and if S is an analytic subset of Y then $f^{-1}(S)$ is an analytic subset of X.

Proof. One may assume that Y is Polish by Lemma 2.4. Then both the graph G of f and X × S are analytic subsets of X × Y by Lemmas 2.4 and 2.5, and so the image of G ∩ (X × S) under the projection of X × Y onto its first coordinate is an analytic subset of X by Proposition 2.6. But this image is just $f^{-1}(S)$. □

The following theorem provides a practical method for showing that subsets of Borel spaces are Borel and is at the same time an important step in the proof of Theorem 2.13.

THEOREM 2.8 (Separation Theorem). If (S_n) is a sequence of mutually disjoint analytic subsets of a countably generated Borel space X then there is a sequence (T_n) of mutually disjoint Borel subsets of X with $S_n \subset T_n$ for each n.

Proof. By Lemma 2.4 and [3, p. 132, Théorème 2] it is enough to show that every analytic subset of a separable metric space is the continuous image of a Polish space. To do this, say that S is an analytic subset of a separable metric space X, and regard X as a subset of its completion Y. There is by definition a Polish space Z, a Borel subset T of Z, and a Borel mapping f of T onto S, and, by [3, p. 127, Proposition 11], a Polish space W and a continuous mapping g of W onto T. It follows easily from Lemma 2.4 that the graph G of f ∘ g is a Borel subset of W × Y and then in turn from [3, p. 127, Proposition 11] that there is a Polish space V and a continuous mapping h of V onto G. Now just notice that S = q(h(V)), where q is the projection of W × Y onto its second coordinate. □

COROLLARY 2.9. If S is a subset of a countably generated Borel space X and if S and X − S are both analytic then S is a Borel subset.

Proof. This follows immediately from Theorem 2.8. □

COROLLARY 2.10. An analytic Borel structure cannot properly contain a countably separated Borel structure.

Proof. Suppose that an analytic Borel structure S on a set X properly contains a countably separated Borel structure S' on X. Let S be a set which is in S but not in S', and let S'' be a countably generated Borel structure on X contained in S'. Then as the identity map from (X,S) to (X,S'') is Borel, S and X − S are both analytic subset of (X,S''). But then S must lie in S'' by Corollary 2.9, a contradiction. □

COROLLARY 2.11. A mapping from an analytic Borel space X into an analytic Borel space Y is Borel if and only if its graph is a Borel subset of X × Y.

Proof. Let f be the mapping in question, let G be its graph, and let p and q be the projections of X × Y onto its first and second coordinates, respectively. It follows easily from Lemma 2.4 that G is a Borel set if f is a Borel map. Conversely, if G is a Borel set then $p_{|G}$ is a Borel isomorphism of G onto X by Corollary 2.10, and thus $f = q \circ (p_{|G})^{-1}$ is a Borel map. □

COROLLARY 2.12. If f is a one-one Borel mapping of an analytic Borel space X into a countably separated Borel space Y then f is a Borel isomorphism of X onto the Borel subspace of Y based on f(X).

Proof. This is an immediate consequence of Corollary 2.10. □

THEOREM 2.13. The range of a one-one Borel mapping of a standard Borel space into a countably separated Borel space is a Borel set.

Proof. Let X be a standard Borel space, let Y be a countably separated Borel space, and let f be a one-one Borel mapping from X to Y. One may, by replacing the given Borel structure on Y by a smaller one and invoking Lemma 2.4, assume that Y is Polish. Then to prove that f(X) is a Borel subset of Y it is enough, by [3, p. 128, Proposition 12 and p. 134, Théorème 3], to find a Polish space V and a continuous one-one mapping g of V onto f(X). This can be done as follows. As X is standard there is, by definition, a Polish space Z, a Borel subset T of Z, and a one-one Borel mapping h of T onto X. Then the graph G of f ∘ h is a Borel subset of Z × Y by Corollary

2.11, and so there will be a Polish space V and a continuous one-one mapping k of V onto G by [3, p. 128, Proposition 12 and p. 134, Théorème 3]. But then g = q ∘ k will do, where q is the projection of Z × Y onto its second coordinate. □

The last two results taken together are analogous to the well-known result in point-set topology which asserts that a continuous injective map between two compact Hausdorff spaces has a closed range and is a homeomorphism of its domain onto its range. After these theorems on Borel spaces the final result of this section is perhaps somewhat of a let down: it implies that standard Borel spaces are classified up to Borel isomorphism by their cardinality and that every uncountable standard Borel space is Borel isomorphic to [0,1].

THEOREM 2.14. Any two uncountable standard Borel spaces are Borel isomorphic.

Proof. Let X be an uncountable standard Borel space. Then X is, by definition, Borel isomorphic to the Borel subspace of a Polish space based on a Borel subset, and then in turn (by Corollary 2.12 and [3, p. 128, Proposition 12 and p. 134, Théorème 3]) to a completely disconnected Polish space Y. Then by [3, p. 143, Exercise 7(c)], Y must contain a countable set D such that Y − D is homeomorphic to $\mathbb{N}^{\mathbb{N}}$. Now let E be an infinite countable subset of Y − D, let φ be a one-one mapping of E onto D ∪ E, and define a function f from Y − D to Y by putting

$$f(x) = \begin{cases} \phi(x) & x \in E \\ x & x \in Y - (D \cup E) \end{cases}$$

Then f is clearly a Borel isomorphism of Y − D onto Y, and thus X must be Borel isomorphic to $\mathbb{N}^{\mathbb{N}}$. □

3. Borel Spaces and Equivalence Relations

Of the four results quoted in this section the first three are certainly the most useful. Roughly speaking, the first theorem says that an equivalence relation on an analytic Borel space is either "very nice" or else "very bad."

THEOREM 3.1. The quotient Borel structure on the quotient space of an analytic Borel space is either analytic or else fails to be countably separated.

Proof. Let ~ be an equivalence relation on an analytic Borel space X, and suppose that the quotient Borel structure S on X/~ is countably separated. Then S contains a countably generated, and hence analytic, Borel structure S'. Now S and S' must be equal, for if not there would be a countably generated Borel structure S'' on X/~ which is contained in S and properly contains S'. But such an S'' would have to be analytic, contradicting Corollary 2.10. □

THEOREM 3.2 (Dixmier's Transversal Theorem). Suppose that ~ is an equivalence relation on a Polish space X satisfying

(a) each equivalence class for ~ is a closed subset of X, and
(b) the saturation for ~ of each open subset of X is a Borel subset of X.

Then there is a Borel transversal for ~.

Proof. See [3, p. 135, Théorème 4]. □

COROLLARY 3.3. If H is a closed subgroup of a second countable locally compact group (or more generally, of a Polish group) G then there is a Borel set in G meeting each left H-coset in exactly one point.

Proof. This follows easily from Proposition 2.1 and Theorem 3.2. □

PROPOSITION 3.4. Let ~ be an equivalence relation on an analytic Borel space X. If there is a Borel cross-section for ~ then the corresponding transversal for ~ is analytic and the quotient Borel structure on X/~ is analytic. If the graph of ~ is analytic and if there is an analytic transversal for ~ then the corresponding cross-section for ~ is Borel.

Proof. The first assertion follows from Theorem 3.1. To prove the second assertion, suppose that the graph G of ~ is analytic and that T is an analytic transversal for ~, and let σ be the corresponding cross-section for ~. Then for an arbitrary Borel set S in X one has

$$\pi^{-1}(\sigma^{-1}(S)) = q([(S \cap T) \times X] \cap G)$$

where q is the projection of X × X onto its second coordinate, and hence $\pi^{-1}(\sigma^{-1}(S))$ is analytic. A similar argument will show that $\pi^{-1}(\sigma^{-1}(X - S))$ too is analytic. Thus $\pi^{-1}(\sigma^{-1}(S))$ is actually a Borel subset of X by Corollary 2.9, and therefore $\sigma^{-1}(S)$ is a Borel subset of X/~. This shows that σ is a Borel map. □

4. Borel Spaces and Measures

The importance of analytic sets and spaces is principally due to three facts: firstly, it is relatively easy to prove that certain sets are analytic; secondly, Corollary 2.9 provides a useful way of showing that certain analytic sets are actually Borel; and thirdly, the results of this section which say, roughly speaking, that analytic sets and spaces are well-behaved measure-theoretically.

It will turn out to be useful to distinguish between functions and equivalence classes of functions. Accordingly, given a Borel measure μ on a Borel space X and a number $p \geq 1$ [respectively, $p = \infty$], let $L^p(\mu)$ denote the set of all those complex valued Borel functions f on X which satisfy $\int_X |f|^p \, d\mu < \infty$ [respectively, which are bounded μ-a.e.], and let $L^p(\mu)$ denote the quotient space obtained from $L^p(\mu)$ by identifying functions which are equal μ-a.e. Then $L^p(\mu)$ is in a natural way a complete seminormed complex linear space and $L^p(\mu)$ is a complex Banach space for $1 \leq p \leq \infty$, and $L^\infty(\mu)$ is even a Banach *-algebra which is *-isomorphic to a von Neumann algebra. Finally, let $f \mapsto f(\mu)$ denote the quotient mapping of $L^p(\mu)$ onto $L^p(\mu)$ (notice that omitting the "p" in this notation does not lead to any ambiguity).

For any Borel space X let M(X) be the Borel space consisting of all of the finite Borel measures on X together with the smallest Borel structure making $\mu \mapsto \int_X f \, d\mu$ a Borel function on M(X) for each bounded real-valued Borel function f on X. If X and Y are two Borel spaces then any Borel map ω from X to Y induces a Borel map ω_* from M(X) to M(Y) which is defined as follows: $\omega_*(\mu)(S) = \mu(\omega^{-1}(S))$ for any μ in M(X) and any Borel set S in Y.

THEOREM 4.1. An analytic subset of a countably generated Borel space is universally measurable.

Proof. Let X be a countably generated Borel space, let S be an analytic subset of X, and let μ be a finite Borel measure on X. By Lemma 2.4 and its proof there is a one-one Borel mapping f of X into a compact metric space Y. Then f(S) is an analytic subset of Y, and so (from the proof of Theorem 2.8) is even the continuous image of a Polish space. But then f(S) is $f_*(\mu)$-measurable by [3, p. 138, Théorème 5], and from this it is trivial to deduce that S is μ-measurable. □

COROLLARY 4.2. Every Borel measure on an analytic Borel space is standard.

Proof. Let μ be a finite Borel measure on an analytic Borel space X. As in the proof of Theorem 4.1, there is a one-one Borel mapping f of X into a Polish space Y. Then f(X), being an analytic subset of Y, is $f_*(\mu)$-measurable by Theorem 4.1, and from this it follows without any difficulty that μ is standard. □

The next result, the so-called *von Neumann-Mackey cross section theorem*, will turn out to be indispensible in the study of direct integrals.

THEOREM 4.3. If X and Y are two analytic Borel spaces, if f is a Borel mapping of X onto Y, and if μ is a Borel measure on Y then there is a μ-null Borel set N in Y and a Borel mapping σ from Y - N to X such that f ∘ σ is the identity on Y - N, or equivalently, there is a Borel mapping σ from Y to X such that f ∘ σ is the identity on some Borel subset of Y whose complement is μ-null.

Proof. This theorem is essentially the same as [41, Theorem 6.3] but is most easily deduced from Hoffmann-Jørgensen's notes [27]. First notice that an analytic Borel space can be given a separable metric topology generating the given Borel structure and making it the continuous image of a Polish space (cf. the proof of Theorem 2.8), and hence that X, Y, and (by Corollary 2.11) the graph G of f are all analytic spaces in the sense of [27]. Then by [27, p. 137, Theorem 6] and the assumption that f is onto there will be a map σ from Y to X such that $(\sigma(y),y) \in G$ for all $y \in Y$ and such that the inverse image by σ of each Borel subset of X is μ-measurable. Thus f ∘ σ is the identity of Y, and an easy argument will show that the restriction of σ to the complement of some μ-null Borel subset of Y is Borel. Indeed, take a sequence of Borel subsets (A_n) of X which generates the given Borel structure. Then there will be sequences (B_n) and (C_n) of Borel subsets of Y such that $B_n \subset \sigma^{-1}(A_n) \subset C_n$ and $\mu(C_n - B_n) = 0$ for each n. Now if $N = \cup_n (C_n - B_n)$ then $\sigma^{-1}(A_n) \cap (Y-N) = B_n \cap (Y-N)$ for each n, and hence the restriction of σ to Y - N is Borel. □

PROPOSITION 4.4. M(X) is standard if X is standard and analytic if X is analytic.

Proof. The result is trivial if X is countable, so assume otherwise. If X is standard one may (by Theorem 2.14) assume that it is the closed unit interval. Then M(X) and the positive cone in the dual of C(X), the continuous complex-valued functions on X, coincide as sets, the Borel structure on M(X) is that generated by the weak *-topology, and this Borel structure is standard by Proposition 2.6. If X is merely analytic there will be a

4. BOREL SPACES AND MEASURES

standard Borel space Y and a Borel function ϕ mapping Y onto X. But then one can use Theorem 4.3 to show that ϕ_* is onto M(X), and so M(X), which is clearly countably generated, is analytic. □

The following theorem is known by various names (the *measure disintegration* or *decomposition* or *fibration theorem*, among others) and has at least as many proofs as it does names; the elegant proof given here is due to E.G. Effros.

THEOREM 4.5. Suppose that X is an analytic Borel space, that Y is a countably separated Borel space, that ω is a Borel mapping of X onto Y, and that μ is a finite Borel measure on X. Then there is a Borel function $\xi \mapsto \mu_\xi$ from Y to M(X) satisfying

(a) $(X - \omega^{-1}(\xi)) = 0$ for each ξ in Y, and

(b) if f and g are bounded real-valued Borel functions on X and Y, respectively, then

$$\int_X f(\zeta)g(\omega(\zeta))d\mu(\zeta) = \int_Y g(\xi) \int_X f(\zeta)d\mu_\xi(\zeta) d\omega_*(\mu)(\xi)$$

Proof. Put $\nu = \omega_*(\mu)$. The result is trivial if X is countable, so assume otherwise. Then (by Corollary 4.2 and Theorem 2.14) one may as well assume that X is Borel isomorphic to $\{0,1\}^{\mathbb{N}}$, and therefore that there is a compact metric topology on X generating the given Borel structure and having as a base a countable field S of sets which are both open and closed. For each set S in S, $T \mapsto \mu(S \cap \omega^{-1}(T))$ is a Borel measure on Y which is absolutely continuous with respect to ν, and hence there is a Borel function g_S on Y satisfying

$$\mu(S \cap \omega^{-1}(T)) = \int_T g_S \, d\nu$$

for all Borel subsets T of Y. These functions must satisfy $0 \leq g_S \leq 1$ ν-a.e. and $g_{S_1 \cup S_2} = g_{S_1} + g_{S_2}$ ν-a.e. for all sets S, S_1 and S_2 in S with $S_1 \cap S_2 = \emptyset$, and hence can be chosen so that $S \mapsto g_S(\xi)$ is a finitely additive probability measure on S for each ξ in Y. But then for each ξ in Y there is a Borel measure μ_ξ on X with $\mu_\xi(S) = g_S(\xi)$ for all S in S (see [25, Section 13, Theorem A]). The monotone class theorem and standard measure-theoretic arguments will show that $\xi \mapsto \mu_\xi$ is a Borel map from Y to M(X) and that (b) holds. Now if T is a countable field of Borel subset of Y separating the points of Y then (using (b)) one can find a ν-null Borel set N in Y such that $\mu_\xi(\omega^{-1}(T)) = 1_T(\xi)$ for all $T \in \mathcal{T}$ and all $\xi \in Y - N$. For any point ξ in $Y - N$ there will be a sequence (T_n) in $\mathcal{T}$ with $\{\xi\} = \cap_n T_n$, thus with $\mu_\xi(T_n) = 1$ for each n, and therefore $\mu_\xi(\omega^{-1}(\xi)) = 1$. Thus the μ_ξ will satisfy (a) provided one redefines μ_ξ to be zero for $\xi \in N$. □

THEOREM 4.6. If μ is a finite Borel measure on a Borel space X, if ν is a Borel measure on an analytic Borel space Y, and if Φ is an identity-preserving weak *-continuous *-homomorphism of $L^\infty(\nu)$ into $L^\infty(\mu)$ then there is a Borel map ω from X to Y such that $\omega_*(\mu) \ll \nu$ and $\Phi(g(\nu)) = (g \circ \omega)(\mu)$ for all $g \in L^\infty(\mu)$. Moreover, if X too is analytic then

(a) Φ is one-one if and only if $\omega_*(\mu) \equiv \nu$, and in this case there is a ν-null Borel set N in Y with $Y - N \subset \omega(X)$,

(b) Φ is onto $L^\infty(\mu)$ if and only if there is a μ-null Borel set M in X such that ω is one-one on $X - M$, and

(c) Φ is one-one and onto $L^\infty(\mu)$ if and only if $\omega_*(\mu) \equiv \nu$ and if there is a μ-null Borel set M in X and a ν-null Borel set N in Y such that ω restricts to a Borel isomorphism of $X - M$ onto $Y - N$.

Proof. The theorem is trivial if Y is countable, so (by Corollary 4.2 and Theorem 2.14) one may assume that $Y = \{0,1\}^\mathbb{N}$ and, in proving (a) - (c), that X is standard. Put $T_k = \{(\xi_n) \in Y: \xi_k = 0\}$ and choose a Borel set S_k in X with $\Phi(1_{T_k}(\nu)) = 1_{S_k}(\mu)$ for each k. Then the function ω from X to Y defined by putting

$$\omega(\zeta)_k = \begin{cases} 0 & \text{if } \xi \in S_k \\ 1 & \text{otherwise} \end{cases}$$

for all points ζ in X has (by the monotone class theorem) all of the required properties. The proof of (a) depends on Theorem 4.1. If Φ is onto, if (S_k) is a sequence of Borel sets in X separating the points of X, and if (T_k) is a sequence of Borel set in Y with $\Phi(1_{T_k}(\nu)) = 1_{S_k}(\mu)$ for each k, then ω is one-one on $X - \cup(S_k \triangle \omega^{-1}(T_k))$; here $\triangle$ denotes symmetric difference. The other half of (b) requires Theorem 2.13, and (c) is now easy. □

HISTORICAL COMMENTS

Although most of the results in what is now known as the theory of Borel spaces were proven between 1917 and 1930 and appear in Kuratowski's book [32], the first edition of which was published in 1933, the theory as one

4. BOREL SPACES AND MEASURES

now knows it was worked out by Mackey in the 1950s [40, 41]. In fact, in reading Mackey's account of the theory one is forced to make a number of excursions into Kuratowski's book. Subsequently, of course, a number of other accounts of the theory have been published by various authors, the most notable (from the point of view of operator algebras and groups representations) being those by Arveson [1], Auslander and Moore [2], and Pedersen [52].

Most of the results of Section 2 can be found in Kuratowski's book [32]. Proposition 2.3 and its proof are taken from [31, p. 278]. Corollary 2.9 is sometimes called Souslin's theorem and is due to Souslin [55, Theorem III], while the more fundamental Theorem 2.8 is due to Lusin [37, p. 52]. Although some authors also refer to Theorem 2.13 as Souslin's theorem, it appears to be due to Lusin [36, Theorem IV; 37, p. 60] in the special case where the spaces in question are subsets of a Euclidean space and the map is continuous; the general case (which is easily reduced to the special case (cf. the proof given here)) seems to be due to Kuratowski [32, p. 397]. The final result of Section 2, viz., Theorem 2.14, also appears to be due to Kuratowski [32, p. 358].

Dixmier's transversal theorem first appeared in [8, Lemme 3]. The original version of the von Neumann-Mackey cross section theorem was proven by von Neumann in connection with his work on direct integrals [49, Lemma 5]; the formulation given here is essentially due to Mackey [41, Theorem 6.3]. Proposition 4.4 is due to Effros [16, p. 444] and Varadarajan [57, Theorem 2.1]. Theorem 4.5 is a folk theorem and has been discussed by a number of authors since the 1940s; the proof given here is taken from [16, Lemma 4.4] (this paper also contains references to other proofs). Finally, Theorem 4.6 is due to von Neumann [48].

Chapter 2
DIRECT INTEGRALS OF HILBERT SPACES AND OPERATORS

5. Hilbert Space-valued L^2-spaces

Consider a (possibly finite-dimensional) separable Hilbert space H. The strong topology on H is by definition Polish, and so generates a standard Borel structure. This is the same Borel structure as is generated by the weak topology, for if (y_j) is a dense sequence in the unit ball of H then

$$\{x \in H: \|x - x_0\| < \epsilon\} = \cap_j \{x \in H: |<x - x_0, y_j>| < \epsilon\}$$

for any $x_0 \in H$ and $\epsilon > 0$. Thus a function v from a Borel space Z to H is Borel if and only if $<v(\cdot),x>$ is a Borel function on Z for each x in H (or equivalently, for each x in some total subset of H). If v and w are two Borel functions from Z to H and if (x_j) is an orthonormal basis for H then

$$<v(\zeta),w(\zeta)> = \sum_j <v(\zeta),x_j><x_j,w(\zeta)>$$

for each point ζ in Z, and thus $\zeta \mapsto <v(\zeta),w(\zeta)>$ is a Borel function on Z. If μ is a Borel measure on Z then the set $L^2(\mu;H)$ consisting of all those Borel functions v from Z to H for which $\|v(\cdot)\|$ is μ-square-integrable is a semi-inner product space with respect to the pointwise linear operations and the semi-inner product

$$<v,w> = \int_Z <v(\zeta),w(\zeta)>d\mu(\zeta)$$

Let $L^2(\mu;H)$ be the associated inner-product space, i.e., $L^2(\mu;H)$ modulo the kernel of the semi-norm derived from the semi-inner product, and let $v \mapsto v(\mu)$ denote the canonical mapping of $L^2(\mu;H)$ onto $L^2(\mu;H)$. If H happens to be the one-dimensional Hilbert space C then $L^2(\mu;H)$ and $L^2(\mu;H)$ are just the usual pre-Hilbert and Hilbert spaces $L^2(\mu)$ and $L^2(\mu)$, respectively. There is an easy way to construct functions (in fact, a total set of functions (see Proposition 5.2)) in $L^2(\mu;H)$: just take a function f in $L^2(\mu)$ and a vector x in H and consider the function fx defined by $(fx)(\zeta) = f(\zeta)x$, $\zeta \in Z$.

THEOREM 5.1. $L^2(\mu;H)$ is a Hilbert space.

Proof. Suppose that (v_n) is a Cauchy sequence in $L^2(\mu;H)$. As (v_n) is convergent provided that some subsequence is convergent, one may assume (by passing, if necessary, to a suitable subsequence) that $\|v_{n+1} - v_n\| < 2^{-n}$ for each n. Then

$$\int_Z \sum_{n=1}^{\infty} \left[\|v_{n+1}(\zeta) - v_n(\zeta)\|^2 \, d\mu(\zeta) \right]^{\frac{1}{2}}$$

$$\leq \sum_{n=1}^{\infty} \left[\int_Z \|v_{n+1}(\zeta) - v_n(\zeta)\|^2 \, d\mu(\zeta) \right]^{\frac{1}{2}}$$

$$< 1$$

by the triangle inequality and Lebesgue's monotone convergence theorem, and so there is a μ-null Borel set N in Z with

$$\sum_{n=1}^{\infty} \|v_{n+1}(\zeta) - v_n(\zeta)\| < \infty \quad \text{for all} \quad \zeta \in Z - N$$

From this it follows easily that $(v_n(\zeta))$ is a Cauchy sequence in H for each point ζ in $Z - N$, and hence that one can define a function v from Z to H by putting

$$v(\zeta) \begin{cases} 0 & \zeta \in N \\ \lim_{n \to \infty} v_n(\zeta) & \zeta \in Z - N \end{cases}$$

Then v is clearly a Borel function and is such that

$$\left[\int_Z \|v(\zeta) - v_m(\zeta)\|^2 \, d\mu(\zeta) \right]^{\frac{1}{2}}$$

$$= \left[\int_Z \lim_{k \to \infty} \|v_k(\zeta) - v_m(\zeta)\|^2 \, d\mu(\zeta) \right]^{\frac{1}{2}}$$

$$\leq \left[\liminf_{k \to \infty} \int_Z \|v_k(\zeta) - v_m(\zeta)\|^2 \, d\mu(\zeta) \right]^{\frac{1}{2}}$$

6. OPERATORS ON HILBERT SPACE-VALUED L^2-SPACES

$$\leq \liminf_{k\to\infty} \sum_{n=m}^{k-1} \|v_{n+1}(\zeta) - v_n(\zeta)\|$$

$$\leq 2^{1-m}$$

for all m by Fatou's lemma. This implies both that v lies in $L^2(\mu;H)$ and that the sequence (v_n) converges to v. □

PROPOSITION 5.2. There is a unique linear isometry of $L^2(\mu;H)$ onto $L^2(\mu) \otimes H$ mapping $(fx)(\mu)$ onto $f(\mu) \otimes x$ for all $f \in L^2(\mu)$ and $x \in H$.

Proof. It is obvious that at most one such isometry can exist. For any $f_1, \ldots, f_n$ in $L^2(\mu)$ and any $x_1, \ldots, x_n$ in H a trivial calculation shows that

$$\|\sum_{j=1}^{n}(f_j x_j)(\mu)\|^2 = \sum_{j=1}^{n}\|f_j(\mu) \otimes x_j\|^2$$

There is therefore a well-defined linear isometry from the closed subspace of $L^2(\mu;H)$ generated by the $(fx)(\mu)$, $f \in L^2(\mu)$ and $x \in H$, onto $L^2(\mu) \otimes H$ with the required property, and so it remains to show that this subspace is all of $L^2(\mu;H)$. Take a sequence (x_k) which is dense in H, and suppose that v is a function in $L^2(\mu;H)$ such that

$$0 = \langle fx_k, v \rangle = \int_Z f(\zeta)\langle x_k, v(\zeta)\rangle d\mu(\zeta)$$

for all f in $L^2(\mu)$ and all k. Then for each k there is a μ-null Borel set N_k in Z with $\langle x_k, v(\cdot)\rangle = 0$ on $Z - N_k$. But then $v = 0$ on $Z - (\cup_k N_k)$, and so $\|v\| = 0$. This shows that the subspace in question is in fact all of $L^2(\mu;H)$. □

COROLLARY 5.3. $L^2(\mu;H)$ is separable whenever μ is standard.

Proof. As $L^2(\mu)$ is separable whenever μ is standard (this can be seen from Theorem 2.14) the corollary is clear in view of the preceeding proposition. □

6. Operators on Hilbert Space-valued L^2-spaces

Let H continue to denote a separable Hilbert space, Z a Borel space, and μ a Borel measure on Z. It is not hard to see that the strong operator topology on $L(H)$ generates the same Borel structure as both the weak and ultrastrong operator topologies, and hence that the weak, ultraweak, strong and ultrastrong operator topologies all generate the same Borel structure on

$L(H)$. In all of what follows $L(H)$ will be considered as carrying five topologies (the four just mentioned together with the norm-topology) and this one Borel structure. Notice that $L(H)$ is then a standard Borel space (this follows immediately from Proposition 2.6 and the compactness in the weak operator topology of the norm-closed balls in $L(H)$) and that a map A from Z to $L(H)$ is Borel if and only if the functions $\langle A(\cdot)x,y\rangle$ (or equivalently, the functions $\langle A(\cdot)x,x\rangle$) are Borel for all vectors x and y in H (or equivalently, for all vectors x and y in some total subset of H). The involution on $L(H)$, being weakly continuous, and the multiplication from $L(H) \times L(H)$ to $L(H)$, being jointly strongly continuous on bounded sets, are Borel maps. Moreover, the map $A \mapsto \|A\|$ on $L(H)$ and the map $(A,x) \mapsto Ax$ from $L(H) \times H$ to H are Borel. Indeed, if (x_j) is a dense sequence in the unit ball of H then $\|A\| = \sup_{j,k} |\langle Ax_j, x_k \rangle|$, and if (y_j) is an orthonormal basis in H then

$$\langle Ax, y \rangle = \sum_j \langle x, y_j \rangle \langle Ay_j, y \rangle$$

for all y in H. This means, in particular, that with respect to the pointwise operations the Borel functions from Z to $L(H)$ form a *-algebra and the Borel functions from Z to H form a module over this algebra. Moreover, if A is a Borel function from Z to $L(H)$ which is μ-essentially bounded (in the sense that $\|A(\cdot)\| \in L^\infty(\mu)$) and if v lies in $L^2(\mu;H)$ then the function Av from Z to H defined by $(Av)(\zeta) = A(\zeta)v(\zeta)$, $\zeta \in Z$, again lies in $L^2(\mu;H)$ and satisfies

$$\|(Av)(\mu)\| \leq \left[\mu - \text{ess.sup} \|A(\cdot)\|\right] \|v(\mu)\|$$

thus A induces a continuous linear operator of norm at most $\mu - \text{ess.sup}\|A(\cdot)\|$, and this operator will be denoted by $A(\mu)$. A constant function and a function of the form $\zeta \mapsto \phi(\zeta)I$ for some $\phi \in L^\infty(\mu)$ are examples of such $L(H)$-valued functions. Let W be the linear isometry of $L^2(\mu;H)$ onto $L^2(\mu) \otimes H$ constructed in Proposition 5.2. It is easy to verify that if A is the constant function from Z to $L(H)$ with constant value A_0 then $WA(\mu)W^{-1} = I \otimes A_0$ and that if ϕ is a function in $L^\infty(\mu)$ then $W[\phi(\cdot)I](\mu)W^{-1} = M_\phi \otimes I$; here M_ϕ is the operator on $L^2(\mu)$ induced by the map $f \mapsto \phi f$ on $L^2(\mu)$.

A *decomposable operator* on $L^2(\mu;H)$ is by definition an operator of the form $A(\mu)$ for some μ-essentially bounded Borel function A from Z to $L(H)$, and a *diagonalizable operator* is one of the form $[\phi(\cdot)I](\mu)$ for some $\phi \in L^\infty(\mu)$. According to the preceeding paragraph, the set of diagonalizable

6. OPERATORS ON HILBERT SPACE-VALUED L^2-SPACES

operators on $L^2(\mu;H)$ is a von Neumann algebra which is *-isomorphic to $\{M_\phi : \phi \in L^\infty(\mu)\}$ and, in turn, to $L^\infty(\mu)$. In Theorem 6.2 the set of decomposable operators will be identified with the commutant of the diagonalizable operators, and so it too is a von Neumann algebra.

PROPOSITION 6.1. (a) $\|A(\mu)\| = \mu\text{-ess.sup}\|A(\cdot)\|$ for any μ-essentially bounded Borel function A from Z to $L(H)$.

   (b) The map $A \mapsto A(\mu)$ is a *-homomorphism from the *-algebra of all μ-essentially bounded Borel functions from Z to $L(H)$ to the *-algebra $L(L^2(\mu;H))$.

   (c) If A and B are two μ-essentially bounded Borel functions from Z to $L(H)$ then $A(\mu) = B(\mu)$ if and only if $A(\cdot) = B(\cdot)$ μ-a.e.

   (d) If A is a μ-essentially bounded Borel function from Z to $L(H)$ then $A(\mu)$ is unitary [respectively, positive, a projection, a partial isometry] if and only if $A(\zeta)$ is unitary [respectively, positive, a projection, a partial isometry] for μ-a.e. $\zeta \in Z$.

Proof. The proofs of (b)-(d) are easy. To prove (a), consider a μ-essentially bounded Borel function A from Z to $L(H)$ and recall that the inequality $\|A(\mu)\| \leq \mu\text{-ess.sup}\|A(\cdot)\|$ has already been established. To show that this is actually an equality, take a sequence (x_j) which is dense in the unit ball of H and notice that

$$\int_S \|A(\zeta)x_j\|^2 \, d\mu(\zeta) = \|A(1_S x_j)\|^2$$
$$= \|A(\mu)(1_S x_j)(\mu)\|^2$$
$$\leq \int_S \|A(\mu)\|^2 \|x_j\|^2 \, d\mu(\zeta)$$

for each Borel set S in Z with $\mu(S) < \infty$. This means that there is a μ-null Borel set N in Z such that $\|A(\zeta)x_j\| \leq \|A(\mu)\| \|x_j\|$ for all j and all $\zeta \in Z - N$, and hence that $\|A(\cdot)\| \leq \|A(\mu)\|$ on $Z - N$. □

THEOREM 6.2. An operator on $L^2(\mu;H)$ is decomposable if and only if it commutes with every diagonalizable operator.

Proof. It is obvious that every decomposable operator commutes with every diagonalizable operator. Conversely, say that T is an operator on $L^2(\mu;H)$ commuting with every diagonalizable operator. Let h be some fixed function in $L^2(\mu)$ whose range is contained in $(0,\infty)$ (if μ is finite h = 1 will of course do) and for each vector x in H choose arbitrarily a function v_x in

$L^2(\mu;H)$ with $v_x(\mu) = T[(hx)(\mu)]$. Then $x \mapsto v_x(\mu)$ is linear, so for any complex numbers $a_1,\ldots,a_n$ and vectors $x_1,\ldots,x_n$ in H it must be the case that

$$a_1 v_{x_1} + \ldots + a_n v_{x_n} = v_{a_1 x_1 + \ldots + a_n x_n} \quad \mu\text{-a.e.}$$

Moreover, for any vector x in H and any function ϕ in $L^\infty(\mu)$ one has

$$\begin{aligned}\int_Z |\phi(\zeta)|^2 \|v_x(\zeta)\|^2 \, d\mu(\zeta) &= \|[\phi(\cdot)I](\mu) v_x(\mu)\|^2 \\ &= \|[\phi(\cdot)I](\mu) T[(hx)(\mu)]\|^2 \\ &= \|T[\phi(\cdot)I](\mu)(hx)(\mu)\|^2 \\ &\leq \|T\|^2 \|[\phi(\cdot)I](\mu)(hx)(\mu)\|^2 \\ &= \|T\|^2 \int_Z |\phi(\zeta)|^2 h(\zeta)^2 \|x\|^2 \, d\mu(\zeta)\end{aligned}$$

and therefore (as ϕ was arbitrary)

$$\|v_x(\cdot)\| \leq \|T\| h(\cdot) \|x\| \quad \mu\text{-a.e.}$$

Now let H_0 be a countable dense subset of H which is a vector space over $Q + iQ$. From the preceeding paragraph and the countability of H_0 it follows that there is a μ-null Borel set N in Z such that for all $\zeta \in Z - N$,

(a) $x \mapsto v_x(\zeta)$ is $(Q + iQ)$-linear from H_0 to H

and

(b) $\|v_x(\zeta)\| \leq \|T\| \|x\| h(\zeta)$ for all $x \in H_0$

Thus for each ζ in $Z - N$ there is a linear operator, say $A(\zeta)$, on H of norm at most $\|T\|$ and satisfying $A(\zeta)x = h(\zeta)^{-1} v_x(\zeta)$, $x \in H_0$. Put $A(\zeta) = 0$ for $\zeta \in N$. Then, by definition, A is a bounded Borel function from Z to $L(H)$ satisfying

$$\begin{aligned}A(\mu)((\phi h)x)(\mu) &= [\phi(\cdot)I](\mu) v_x(\mu) \\ &= [\phi(\cdot)I](\mu) T[(hx)(\mu)] \\ &= T[\phi(\cdot)I](\mu)(hx)(\mu) \\ &= T((\phi h)x)(\mu)\end{aligned}$$

for all $\phi \in L^\infty(\mu)$ and all $x \in H_0$. Thus $A(\mu)$ and T agree on a subset of $L^2(\mu;H)$ which (by Proposition 5.2) is total, and so $A(\mu) = T$. $\square$

6. OPERATORS ON HILBERT SPACE-VALUED L^2-SPACES

The following rather technical result will be used on a number of occasions in the following chapters.

PROPOSITION 6.3. Let $A, A_1, A_2, \ldots$ be μ-essentially bounded Borel functions from Z to $L(H)$.

(a) If $A_n(\mu) \to A(\mu)$ strongly as $n \to \infty$ then there is a subsequence (n_k) such that $A_{n_k}(\zeta) \to A(\zeta)$ strongly as $k \to \infty$ for μ-a.a. $\zeta \in Z$.

(b) If $\sup_n \|A_n(\zeta)\| < \infty$ and if $A_n(\zeta) \to A(\zeta)$ strongly as $n \to \infty$ for μ-a.a. $\zeta \in Z$ then $A_n(\mu) \to A(\mu)$ strongly as $n \to \infty$.

Proof. One may as well suppose that A is the zero function. The hypothesis in part (a) implies that $\sup_n \|A_n(\mu)\| < \infty$, so under the hypothesis of either part of the proposition there will be (by Proposition 6.1) a positive number a and a μ-null Borel set M in Z with $\|A_n(\mu)\| \le a$ for all n and all $\zeta \in Z - M$.

(a) Let h be as in the proof of Theorem 6.2 and let (x_j) be a dense sequence in H. Then

$$\int_Z h(\zeta)^2 \|A_n(\zeta)x_j\|^2 \, d\mu(\zeta) = \|A_n(\mu)(hx_j)(\mu)\|^2 \to 0$$

as $n \to \infty$ for each fixed j. Taking j = 1, there will thus be a subsequence $(n_{1,k})$ and a μ-null Borel set N_1 in Z such that $\|A_{n_{1,k}}(\zeta)x_1\| \to 0$ as $k \to \infty$ for all $\zeta \in Z - N_1$. Now taking j = 2, there will be a subsequence $(n_{2,k})$ of $(n_{1,k})$ and a μ-null Borel set N_2 in Z such that $\|A_{n_{2,k}}(\zeta)x_2\| \to 0$ as $k \to \infty$ for all $\zeta \in Z - N_2$. Continuing in this way and letting $n_k = n_{k,k}$ and $N = \cup_k N_k$, one obtains a subsequence (n_k) such that $\|A_{n_k}(\zeta)x_j\| \to 0$ as $k \to \infty$ for all j and all $\zeta \in Z - N$. But then $A_{n_k}(\zeta) \to 0$ strongly as $k \to \infty$ for all $\zeta \in Z - (M \cup N)$.

(b) Consider a function f in $L^2(\mu) \cap L^\infty(\mu)$ and a vector x in H. Then

$$\|A_n(\mu)(fx)(\mu)\|^2 = \int_Z |f(\zeta)|^2 \|A_n(\zeta)x\|^2 \, d\mu(\zeta)$$

and for μ-a.a. $\zeta \in Z$ it is the case that $\lim_{n\to\infty} \|A_n(\zeta)x\| = 0$ and

$$|f(\zeta)| \, \|A_n(\zeta)x\| \le a\|x\| \left[\mu - \text{ess.sup}|f(\cdot)|\right]$$

and hence $\|A_n(\mu)(fx)(\mu)\| \to 0$ as $n \to \infty$ by the Lebesgue dominated convergence theorem. From this and Proposition 5.2 it is clear that $A_n(\mu) \to 0$ strongly as $n \to \infty$. □

7. Fields of Hilbert Spaces and Operators

The Hilbert space $L^2(\mu;H)$ constructed in Section 5 can be thought of as the integral with respect to μ of the Hilbert space-valued function on Z which has constant value H. In this section non-constant Hilbert space-valued functions will be integrated to yield the so-called "direct integral" Hilbert spaces, and, moreover, the results of the preceeding section will be generalized to such Hilbert spaces.

In order to define these "direct integral" Hilbert spaces it will be necessary to be able to compare any separable Hilbert space of a given dimension with a standard, or distinguished, Hilbert space of that dimension. How these standard Hilbert spaces are defined is actually immaterial, although it will be convenient to have them nested; what is important is that they be fixed once and for all. Perhaps the easiest way to define them is as follows: let $\ell^2 = \ell_\infty^2$ be the Hilbert space of all square-summable sequences and for each non-negative integer n let ℓ_n^2 consist of those elements of ℓ^2 whose values at n + 1, n + 2,... are zero. Then $\ell_0^2 \subset \ell_1^2 \subset ... \subset \ell_\infty^2$ and dim ℓ_n^2 = n for $n \in \mathbb{N} \cup \{\infty\}$.

Suppose for the moment that Z is merely a set. A *field of Hilbert spaces* on Z is by definition a Hilbert space-valued function on Z, i.e., a rule which assigns to each point in Z a Hilbert space. If H is a field of Hilbert spaces on Z then, with respect to the pointwise operations, $\Pi_{\zeta \in Z} H(\zeta)$ is a vector space and $\Pi_{\zeta \in Z} L(H(\zeta))$ is a *-algebra; elements of the former are called *vector fields over H*, and of the latter, *operator fields over H*. A *coherence* α for a field of Hilbert spaces H on Z is a choice, for each point ζ in Z, of a linear isometry $\alpha(\zeta)$ of $H(\zeta)$ into ℓ^2 whose range is $\ell_{\dim H(\zeta)}^2$. (It might appear more natural to regard $\alpha(\zeta)$ as a linear isometry of $H(\zeta)$ onto $\ell_{\dim H(\zeta)}^2$, but this would lead to the technical complication that the domain of $\alpha(\zeta)^*$ would be $\ell_{\dim H(\zeta)}^2$, and so would depend on ζ.) It is clear that a field of Hilbert spaces H on Z has a coherence if and only if $H(\zeta)$ is separable for each $\zeta \in Z$. Finally, a *constant field of Hilbert spaces on Z* is a field of Hilbert spaces H on Z of the form $H(\zeta) = H_0$ for all $\zeta \in Z$ and some fixed Hilbert space H_0, and a *constant coherence* for such an H is a coherence α for H such that $\alpha(\zeta) = \theta$ for all $\zeta \in Z$ and some fixed linear isometry θ of H_0 into ℓ^2.

Now suppose that Z is a Borel space and that α is a coherence for a field of Hilbert spaces H on Z. Put $Z_n = \{\zeta \in Z: \dim H(\zeta) = n\}$ for $n = \infty, 0, 1, 2, \ldots$ and call H a *Borel field of Hilbert spaces* if each of these sets is Borel. (It will turn out that virtually all of the fields of Hilbert

7. FIELDS OF HILBERT SPACES AND OPERATORS

spaces to be considered in these notes will be Borel in this sense.) An α-*Borel vector* [respectively, *operator*] *field over* H is a vector field v [respectively, an operator field A] over H for which the map $\zeta \mapsto \alpha(\zeta)v(\zeta)$ [respectively, $\zeta \mapsto \alpha(\zeta)A(\zeta)\alpha(\zeta)*$] from Z to ℓ^2 [respectively, to $L(\ell^2)$] is Borel. Under the pointwise operations the α-Borel vector fields over H form a vector space and the α-Borel operator fields over H form a *-algebra. Moreover, it follows from Sections 5 and 6 that if v and w are two α-Borel vector fields over H and A is an α-Borel operator field over H then $\zeta \mapsto A(\zeta)v(\zeta)$ is again an α-Borel vector field over H and that both $\zeta \mapsto <v(\zeta),w(\zeta)>$ and $\zeta \mapsto \|A(\zeta)\|$ are Borel functions on Z. If H is Borel then a vector field v [respectively, an operator field A] over H is evidently α-Borel if and only if the map $\zeta \mapsto \alpha(\zeta)v(\zeta)$ [respectively, $\zeta \mapsto \alpha(\zeta)A(\zeta)\alpha(\zeta)*|\ell_n^2$] from Z_n to ℓ_n^2 [respectively, to $L(\ell_n^2)$] is Borel (with respect to the relative Borel structure on Z_n) for each n.

Finally, suppose that μ is a Borel measure on a Borel space Z and that α is a coherence for a Borel field of Hilbert spaces H on Z. Then the set $L^2(\mu;H,\alpha)$ consisting of all those α-Borel vector fields v over H for which $\|v(\cdot)\|$ belongs to $L^2(\mu)$ is a semi-inner product space with respect to the pointwise linear operations and the semi-inner product

$$<v,w> = \int_Z <v(\zeta),w(\zeta)> d\mu(\zeta)$$

The symbols $L^2(\mu;H,\alpha)$ and $\int_Z^\alpha H(\zeta)d\mu(\zeta)$ will be used interchangeably to denote the associated inner-product space (which is actually a Hilbert space (see Theorem 7.1)) and $v \mapsto \int_Z^\alpha v(\zeta)d\mu(\zeta)$ the canonical mapping of $L^2(\mu;H,\alpha)$ onto $L^2(\mu;H,\alpha)$. Just as in Section 6, an α-Borel μ-essentially bounded operator field A over H (where A being μ-essentially bounded again means that $\|A(\cdot)\| \in L^\infty(\mu)$) induces an operator $\int_Z^\alpha A(\zeta)d\mu(\zeta)$ on $L^2(\mu;H,\alpha)$ satisfying

$$\int_Z^\alpha A(\zeta)d\mu(\zeta) \int_Z^\alpha v(\zeta)d\mu(\zeta) = \int_Z^\alpha A(\zeta)v(\zeta)d\mu(\zeta)$$

for all $v \in L^2(\mu;H,\alpha)$; operators of this form are by definition the *decomposable operators* on $L^2(\mu;H,\alpha)$. The *diagonalizable operators* on $L^2(\mu;H,\alpha)$ are the decomposable operators of the form $\int_Z^\alpha \phi(\zeta)I_{H(\zeta)}d\mu(\zeta)$ for some $\phi \in L^\infty(\mu)$.

At this point the reader may find the following trivial example to be somewhat illuminating. Let Z be a countable set with the discrete Borel structure, let μ be counting measure on Z, let H be a field of Hilbert spaces on Z, and let α be a coherence for H. Then $L^2(\mu;H,\alpha)$ is equal to (and not merely isomrophic to) $\oplus_{\zeta \in Z} H(\zeta)$ and the decomposable and diagonalizable operators on $L^2(\mu;H,\alpha)$ are just the elements of $\oplus_{\zeta \in Z} L(H(\zeta))$ and $\oplus_{\zeta \in Z} C(H(\zeta))$, respectively.

Now consider, for a moment, the case in which H is a constant field, say $H(\zeta) = H_0$ for all $\zeta \in Z$, and α is a constant coherence. Then clearly a function v from Z to H_0 as well as a function A from Z to $L(H_0)$ is Borel if and only if it is α-Borel. This means that $L^2(\mu;H,\alpha)$ and $L^2(\mu;H,\alpha) = \int_Z^\alpha H(\zeta)d\mu(\zeta)$ are equal to $L^2(\mu;H_0)$ and $L^2(\mu;H_0)$, respectively, that $\int_Z^\alpha v(\zeta)d\mu(\zeta) = v(\mu)$ and $\int_Z^\alpha A(\zeta)d\mu(\zeta) = A(\mu)$ for all functions v in $L^2(\mu;H)$ and all μ-essentially bounded Borel functions A from Z to $L(H)$. In this case, then, the choice of the constant coherence α is immaterial.

The notation of the form "$\int_Z^\alpha$" which has just been introduced is admittedly somewhat cumbersome, but is at the same time suggestive and close to the traditional notation. It will even be useful to employ this notation in the situation discussed in Sections 5 and 6 but with the coherence suppressed. Namely, if H_0 is a separable Hilbert space, if $H(\zeta) = H_0$ for each $\zeta \in Z$, if v is a function in $L^2(\mu;H_0)$, and if A is a μ-essentially bounded Borel function from Z to $L(H_0)$ then the symbols $\int_Z^\oplus H(\zeta)d\mu(\zeta)$, $\int_Z^\oplus v(\zeta)d\mu(\zeta)$, and $\int_Z^\oplus A(\zeta)d\mu(\zeta)$ will sometimes be written in place of $L^2(\mu;H)$, $v(\mu)$, and $A(\mu)$, respectively.

This is a close connection between the linear spaces and operators just defined and those studied in Sections 5 and 6. To explain this, fix a Borel space Z, a Borel measure μ on Z, a Borel field of Hilbert spaces H on Z, and a coherence α for H. Then for $n = \infty, 0, 1, \ldots$ the set $Z_n = \{\zeta \in Z : \dim H(\zeta) = n\}$ is Borel and one can give it the relative Borel structure and consider the Borel measure μ_n on Z_n which is the restriction of μ to the Borel subsets of Z_n. Then for any v in $L^2(\mu;H,\alpha)$ one has

$$\int_Z \|v(\zeta)\|^2 d\mu(\zeta) = \sum_{0 \leq n \leq \infty} \int_{Z_n} \|\alpha(\zeta)v(\zeta)\|^2 d\mu_n(\zeta)$$

by Lebesgue's monotone convergence theorem. So for each n and each v in $L^2(\mu;H,\alpha)$ the restriction of the map $\zeta \mapsto \alpha(\zeta)v(\zeta)$ to Z_n lies in $L^2(\mu_n;\ell_n^2)$ and there is a linear isometry W from $L^2(\mu;H,\alpha)$ into $\oplus_{0 \leq n \leq \infty} L^2(\mu_n;\ell_n^2)$ satisfying

$$W \int_Z^\alpha v(\zeta)d\mu(\zeta) = \oplus_{0 \leq n \leq \infty} \int_{Z_n}^\oplus \alpha(\zeta)v(\zeta)d\mu(\zeta)$$

for all $v \in L^2(\mu;H,\alpha)$. If w is an element of $\oplus_{0 \leq n \leq \infty} L^2(\mu_n;\ell_n^2)$ then for

7. FIELDS OF HILBERT SPACES AND OPERATORS

each n there will be a function w_n in $L^2(\mu_n;\ell_n^2)$ with $w = \oplus_{0 \leq n \leq \infty} w_n(\mu_n)$. Putting $v(\zeta) = \alpha(\zeta)^* w_n(\zeta)$ for $\zeta \in Z_n$, one obtains an element $\overline{v}$ of $L^2(\mu;H,\alpha)$ satisfying $W \int_Z^\alpha v(\zeta)d\mu(\zeta) = w$. This shows that W is actually onto $\oplus_{0 \leq n \leq \infty} L^2(\mu_n;\ell_n^2)$. If A is a μ-essentially bounded α-Borel operator field over H then an easy calculation will show that

$$W\left[\int_Z^\alpha A(\zeta)d\mu(\zeta)\right]W^{-1} = \oplus_{0 \leq n \leq \infty} \int_{Z_n}^\oplus A_n(\zeta)d\mu_n(\zeta) \qquad (*)$$

where $A_n(\zeta) = \alpha(\zeta)A(\zeta)\alpha(\zeta)^*|\ell_n^2$ for $\zeta \in Z_n$. Conversely, if, for each n, A_n is a μ-essentially bounded Borel function from Z_n to $L(\ell_n^2)$ and if $\sup_n \|A_n(\mu_n)\| < \infty$ then the formula $A(\zeta) = \alpha(\zeta)^* A_n(\zeta)\alpha(\zeta)$, $\zeta \in Z_n$, defines a μ-essentially bounded α-Borel operator field A over H satisfying $(*)$ (use Proposition 6.1). Moreover, in $(*)$ the operator $\int_Z^\alpha A(\zeta)d\mu(\zeta)$ is diagonalizable if and only if each of the operators $\int_{Z_n}^\oplus A_n(\zeta)d\mu_n(\zeta)$ is diagonalizable. This means that if M and N denote the algebras of diagonalizable and decomposable operators on $L^2(\mu;H,\alpha)$, respectively, and M_n the algebra of diagonalizable operators on $L^2(\mu_n;\ell_n^2)$ then

$$M' = \left(W^{-1}\left[\oplus_{0 \leq n \leq \infty} M_n\right]W\right)' = W^{-1}\left[\oplus_{0 \leq n \leq \infty} M_n'\right]W = N$$

by Theorem 6.2. The following theorem now follows easily from the preceeding discussion and Sections 5 and 6.

THEOREM 7.1. If Z,μ,H,α,M and N are as in the preceeding paragraph then

(i) $L^2(\mu;H,\alpha)$ is a Hilbert space which is separable whenever μ is standard,

(ii) there is a sequence (v_n) in $L^2(\mu;H,\alpha)$ with the property that the sequence $(v_n(\zeta))$ is dense in $H(\zeta)$ for each $\zeta \in Z$,

(iii) $\left\|\int_Z^\alpha A(\zeta)d\mu(\zeta)\right\| = \mu\text{-ess.sup}\|A(\cdot)\|$ for any μ-essentially bounded α-Borel operator field A over H,

(iv) $A \mapsto \int_Z^\alpha A(\zeta)d\mu(\zeta)$ is a *-homomorphism from the *-algebra of all μ-essentially bounded α-Borel operator fields over H to $L(L^2(\mu;H,\zeta))$,

(v) M and N are von Neumann algebras on $L^2(\mu;H,\alpha)$, $M = N'$, and $\phi \mapsto \int_Z^\infty \phi(\zeta) I_{H(\zeta)} d\mu(\zeta)$ is a *-homomorphism of $L^\infty(\mu)$ onto N which induces a *-isomorphism of $L^\infty(\mu)$ onto N,

(vi) if $A,A_1,A_2,\ldots$ are μ-essentially bounded α-Borel operator fields over H and if $\int_Z^\alpha A_n(\zeta)d\mu(\zeta) \to \int_Z^\alpha A(\zeta)d\mu(\zeta)$ strongly as $n \to \infty$ then there is a subsequence (n_k) such that $A_{n_k}(\zeta) \to A(\zeta)$ strongly as $k \to \infty$ for μ-a.a. $\zeta \in Z$, and

(vii) if $A, A_1, A_2, \ldots$ are μ-essentially bounded α-Borel operator fields over H, if $\sup_n \|A_n(\mu)\| < \infty$, and if $A_n(\zeta) \to A(\zeta)$ strongly as $n \to \infty$ for μ-a.a. $\zeta \in Z$ then $\int_Z^\alpha A_n(\zeta)d\mu(\zeta) \to \int_Z^\alpha A(\zeta)d\mu(\zeta)$ strongly as $n \to \infty$.

If Z, μ, H and α are as in Theorem 7.1 the Hilbert space $L^2(\mu; H, \alpha)$ will be called the *direct integral Hilbert space* of H with respect to μ and α. The following three results describe some simple but useful technical properties of direct integral Hilbert spaces.

LEMMA 7.2. Suppose that μ and ν are two equivalent Borel measures on a Borel space Z, that H is a Borel field of Hilbert spaces on Z, and that α is a coherence for H. Then there is a Borel function θ from Z to $(0, \infty)$ with $\mu = \theta\nu$ and such that the map $v \mapsto \theta^{\frac{1}{2}} v$ is a linear isometry of $L^2(\mu; H, \alpha)$ onto $L^2(\nu; H, \alpha)$.

Proof. The existence of a function θ satisfying the first condition is just the statement of the Radon-Nikodym theorem, and a trivial calculation will show that this function satisfies the second condition. □

LEMMA 7.3. Let μ be a Borel measure on a Borel space Z and let α be a coherence for a Borel field of Hilbert spaces H on Z. Suppose that (ϕ_j) is a weak *-total family in $L^\infty(\mu)$, and suppose further that (v_n) is a sequence in $L^2(\mu; H, \alpha)$ with the property that the sequence $(v_n(\zeta))$ is total in $H(\zeta)$ for μ-a.a. $\zeta \in Z$. Then the family $(\phi_j v_n)$ is total in $L^2(\mu; H, \alpha)$.

Proof. Say that v is a function in $L^2(\mu; H, \alpha)$ which is orthogonal to $\phi_j v_n$, i.e., that

$$0 = \langle \phi_j v_n, v \rangle = \int_Z \phi_j(\zeta) \langle v_n(\zeta), v(\zeta) \rangle \, d\mu(\zeta)$$

for all j and n. Then for each n the function $\zeta \mapsto \langle v_n(\zeta), v(\zeta) \rangle$, which belongs to $L^2(\mu)$, must be zero off some μ-null Borel set N_n. But then $\langle v_n(\zeta), v(\zeta) \rangle = 0$ for all n and all ζ in $Z - \cup_n N_n$, and hence $v = 0$ μ-a.e. □

LEMMA 7.4. Let Z, μ, H and α be as in Lemma 7.3 and let H_0 be a separable Hilbert space of dimension p. For each element n of $\mathbb{N} \cup \{\infty\}$ choose a linear

8. THE CONSTRUCTION OF COHERENCES

isometry θ_n of $\ell_n^2 \otimes H_0$ onto ℓ_{np}^2 and put

$$\beta(\zeta) = \theta_{\dim H(\zeta)} \circ (\alpha(\zeta) \otimes I), \quad \zeta \in Z$$

Then β is a coherence for the Borel field of Hilbert spaces $H(\cdot) \otimes H_0$ on Z and there is a unique linear isometry of $L^2(\mu;H(\cdot) \otimes H_0, \beta)$ onto $L^2(\mu;H,\alpha) \otimes H_0$ carrying $\int_Z^\beta v(\zeta) \otimes x \, d\mu(\zeta)$ onto $\int_Z^\alpha v(\zeta) d\mu(\zeta) \otimes x$ for all $v \in L^2(\mu;H,\alpha)$ and all $x \in H_0$.

Proof. That there can be at most one such linear isometry is obvious. To actually construct such an isometry, let (e_j) be an orthonormal basis in H_0 and consider a function w in $L^2(\mu;H(\cdot) \otimes H_0, \beta)$. Then for each point ζ in Z there is a uniquely-determined family $(w_j(\zeta))$ in $H(\zeta)$ with $w(\zeta) = \sum_j w_j(\zeta) \otimes e_j$, and these families must be such that $\sum_j \|w_j(\zeta)\|^2 < \infty$ and

$$\langle \alpha(\zeta) w_j(\zeta), x \rangle = \langle \beta(\zeta) w(\zeta), \theta_{\dim H(\zeta)} (x \otimes e_j) \rangle$$

for each $\zeta \in Z$ and $x \in \ell^2$ and such that

$$\int_Z \|w(\zeta)\|^2 \, d\mu(\zeta) = \sum_j \int_Z \|w_j(\zeta)\|^2 \, d\mu(\zeta)$$

(use Lebesgue's monotone convergence theorem). This shows that each of the functions w_j belongs to $L^2(\mu;H,\alpha)$ and that the sum $\sum_j \int_Z^\alpha w_j(\zeta) d\mu(\zeta) \otimes e_j$ converges in $L^2(\mu;H,\alpha) \otimes H_0$ and has the same norm as $\int_Z^\beta w(\zeta) d\mu(\zeta)$. It should now be clear that the desired isometry exists. □

8. The Construction of Coherences

Suppose that one is given a Borel space Z, a Borel measure μ on Z, and a Borel field of Hilbert spaces H on Z. In order to form a direct integral Hilbert space from this data one needs to know which of the vector fields over H are Borel, or more precisely, one needs a coherence. The object of this section is to describe a useful method of constructing coherences, and hence direct integrals. The method can be summarized by remarking that a coherence, being a family of linear isometries, is determined by orthonormal basis and that orthonormal basis, in turn, are determined by total sequences through the Gram-Schmidt orthonormalization process.

Before embarking on the construction it will be convenient to make a definition. Consider a Borel field of Hilbert spaces H on a Borel space Z. Two coherences α and β for H will be called *equivalent* if a vector field over H is α-Borel if and only if it is β-Borel. It is not difficult to see that there is no need to distinguish between equivalent coherences for H. Indeed, suppose that α and β are two equivalent coherences for H. Then the two spaces $L^2(\mu;H,\alpha)$ and $L^2(\mu;H,\beta)$ actually coincide for any Borel measure μ on Z. Suppose that A is an α-Borel operator field over H, and let x and y be two vectors in ℓ^2. Then $\beta(\cdot)^*x$ and $\beta(\cdot)^*y$ are β-Borel, and hence α-Borel, vector fields over H, and therefore

$$\zeta \mapsto \langle A(\zeta)\beta(\zeta)^*x, \beta(\zeta)^*y \rangle = \langle \beta(\zeta)A(\zeta)\beta(\zeta)^*x, y \rangle$$

is a Borel function on Z. This shows that $\zeta \mapsto \beta(\zeta)A(\zeta)\beta(\zeta)^*$ is a Borel map from Z to $L(\ell^2)$, and hence that A is β-Borel.

Now consider a set Z, a field of Hilbert spaces H on Z, and a sequence of vector fields (v_n) over H with the property that for each ζ in Z the sequence $(v_n(\zeta))$ is total in $H(\zeta)$. In applying the Gram-Schmidt orthonormalization process pointwise to this sequence the following notational device will be convenient: if x is a vector in a Hilbert space, put $[x] = 0$ if $x = 0$ and $[x] = \|x\|^{-1}x$ otherwise. Begin, as usual, by inductively defining a new sequence of vector fields (u_n) over H as follows:

$$u_1(\zeta) = [v_1(\zeta)]$$
$$u_2(\zeta) = [v_2(\zeta) - \langle v_2(\zeta), u_1(\zeta) \rangle u_1(\zeta)]$$
$$\vdots$$
$$u_{n+1}(\zeta) = [v_{n+1}(\zeta) - \sum_{k=1}^{n} \langle v_{n+1}(\zeta), u_k(\zeta) \rangle u_k(\zeta)]$$
$$\vdots$$

Then for each point ζ in Z the nonzero members of the sequence $(u_n(\zeta))$ form an orthonormal basis in $H(\zeta)$. Let α be the unique coherence for H with the property that for each point ζ in Z and each positive integer k with $k \leq \dim H(\zeta)$, $\alpha(\zeta)$ maps the kth nonzero member of the sequence $(u_n(\zeta))$ onto e_k, where $e_1, e_2, \ldots$ is the standard orthonormal basis for ℓ^2; thus

$$\alpha(\zeta)u_m(\zeta) = e_k \iff \|u_m(\zeta)\| = 1 \text{ and } \sum_{n=1}^{m} \|u_n(\zeta)\| = k \qquad (1)$$

8. THE CONSTRUCTION OF COHERENCES

Now suppose that, in addition, Z is a Borel space and that $\zeta \mapsto \langle v_m(\zeta), v_n(\zeta) \rangle$ is a Borel function on Z for each m and n. Then there are complex-valued Borel functions $f_{n,m}$ and $g_{n,m}$ on Z, $1 \leq m \leq n < \infty$, such that

$$u_n(\zeta) = \sum_{m=1}^{n} f_{n,m}(\zeta) v_m(\zeta) \qquad (2)$$

and

$$v_n(\zeta) = \sum_{m=1}^{n} g_{n,m}(\zeta) u_m(\zeta) \qquad (3)$$

Thus H is actually a Borel field of Hilbert spaces since $\dim H(\zeta) = p$ if and only if $\sum_{n=1}^{\infty} \|u_n(\zeta)\| = p$, $p = \infty, 0, 1, 2, \ldots$. Moreover, the sets

$$Z(k,m) = \{\zeta \in Z : \alpha(\zeta) u_m(\zeta) = e_k\}, \quad 1 \leq k \leq m < \infty$$

are all Borel by (1), and hence each of the vector fields v_n is α-Borel since

$$\langle \alpha(\zeta) v_n(\zeta), x \rangle = \sum_{m=1}^{n} g_{n,m}(\zeta) \langle \alpha(\zeta) u_m(\zeta), x \rangle$$
$$= \sum_{m=1}^{n} \sum_{k=1}^{m} g_{n,m}(\zeta) 1_{Z(k,m)}(\zeta) \langle e_k, x \rangle$$

for any x in ℓ^2.

PROPOSITION 8.1. Let H be a field of Hilbert spaces on a Borel space Z, and suppose that V is a countable set of vector fields over H such that

  (a) for each ζ in Z the family $(v(\zeta))_{v \in V}$ is total in $H(\zeta)$, and
  (b) $\zeta \mapsto \langle v(\zeta), w(\zeta) \rangle$ is a Borel function on Z for all $v, w \in V$.

Then H is a Borel field and to within equivalence there is a unique coherence α for H making each v in V an α-Borel vector field.

Proof. That H is a Borel field and that there is a coherence α for H making each v in V an α-Borel vector field has already been proven. That there is to within equivalence just one such α comes from the following lemma.

LEMMA 8.2. Let Z, H, and V be as in Proposition 8.1, and let β be a coherence for H making each v in V a β-Borel vector field. Then a necessary and sufficient condition that a vector field w over H be β-Borel is that $\zeta \mapsto \langle w(\zeta), v(\zeta) \rangle$ be a Borel function on Z for each v in V.

Proof. The necessity of the condition is obvious. To prove sufficiency, suppose that w is a vector field over H such that $\zeta \mapsto \langle w(\zeta), v(\zeta) \rangle$ is a Borel function on Z for each v in V. Arrange the elements of V in a sequence (v_n) and let (u_n) be the sequence of vector fields over H constructed from the sequence (v_n) as above. Then

$$\langle \beta(\zeta) w(\zeta), x \rangle = \sum_{n=1}^{\infty} \langle w(\zeta), u_n(\zeta) \rangle \langle \beta(\zeta) u_n(\zeta), x \rangle$$

is a Borel function of ζ on Z for any x in ℓ^2 by (2), and so w is β-Borel. □

9. Direct Integral Decompositions

Suppose that μ is a Borel measure on a standard Borel space Z, that H is a Borel field of Hilbert spaces on Z, and that α is a coherence for H. Then one can form the direct integral Hilbert space $L^2(\mu; H, \alpha)$ and the abelian von Neumann algebra of diagonalizable operators on $L^2(\mu; H, \alpha)$. The present section is concerned with what might be thought of as the converse of this construction. Namely, is every abelian von Neumann algebra M acting on a separable Hilbert space spatially isomorphic to the algebra of diagonalizable operators on some direct integral Hilbert space and, if so, to what extent does M determine the base space, the measure, the field of Hilbert spaces, and the coherence of the direct integral Hilbert space?

THEOREM 9.1. Suppose that M is an abelian von Neumann algebra acting on a separable Hilbert space K. Then there is a standard Borel space Z and a Borel measure μ on Z such that M and $L^\infty(\mu)$ are *-isomorphic, and there is a Borel field of Hilbert spaces H on Z, a coherence α for H, and a linear isometry W of K onto $L^2(\mu; H, \alpha)$ such that WMW^{-1} is the algebra of diagonalizable operators on $L^2(\mu; H, \alpha)$. Moreover, if $\phi \mapsto T_\phi$ is the *-homomorphism of $L^\infty(\mu)$ onto M obtained by composing the canonical mapping of $L^\infty(\mu)$ onto $L^\infty(\mu)$ with some *-isomorphism of $L^\infty(\mu)$ onto M then W can be chosen so that

$$WT_\phi W^{-1} = \int_Z^\alpha \phi(\zeta) I_{H(\zeta)} d\mu(\zeta) \quad \text{for all} \quad \phi \in L^\infty(\mu)$$

Proof. The existence of a standard Borel space Z and a Borel measure μ on Z such that $L^\infty(\mu)$ and M are *-isomorphic is just the statement of Proposition A.4. Fix such a *-isomorphism and let $\phi \mapsto T_\phi$ be the corresponding *-homo-

9. DIRECT INTEGRAL DECOMPOSITIONS

morphism of $L^\infty(\mu)$ onto M. For each pair of vectors x and y in K and each Borel subset S of Z put $\mu_{x,y}(S) = \langle T_{1_S} x, y \rangle$. Then $\mu_{x,y}$ is a complex-valued Borel measure on Z, for if (S_n) is a sequence of mutually disjoint Borel subsets of Z and if $S = \cup_n S_n$ then $(T_{1_{S_n}})$ is a sequence of mutually orthogonal projections in M, $T_{1_S} = \sum_n T_{1_{S_n}}$, and

$$\mu_{x,y}(S) = \langle T_{1_S} x, y \rangle = \sum_n \langle T_{1_{S_n}} x, y \rangle = \sum_n \mu_{x,y}(S_n)$$

Each $\mu_{x,y}$ is clearly absolutely continuous with respect to μ, and so for each pair of vectors x and y in K there is a complex-valued Borel function $h_{x,y}$ on Z with $\mu_{x,y} = h_{x,y}\mu$. Now for any three vectors x, y and z in K and any complex number a one has

$$\mu_{x,y} = \bar{\mu}_{y,x}, \quad \mu_{x+y,z} = \mu_{x,z} + \mu_{y,z}, \quad \text{and} \quad \mu_{ax,y} = a\mu_{x,y}$$

and hence each the three relations

$$h_{x,y} = \bar{h}_{y,x}, \quad h_{x+y,z} = h_{x,z} + h_{y,z'}, \quad \text{and} \quad h_{ax,y} = ah_{x,y} \qquad (*)$$

must hold μ-a.e.

Now let K_0 be a countable dense subset of K which is a vector space over $Q + iQ$. By modifying the functions $h_{x,y}$, $x, y \in K_0$, on a μ-null Borel set if necessary, one may assume that $(*)$ holds everywhere on Z for all choices of vectors x, y, and z in K_0 and all complex numbers a in $Q + iQ$. So for a fixed point ζ in Z the map $(x,y) \mapsto h_{x-y,x-y}(\zeta)^{\frac{1}{2}}$ is a semi-metric, say d_ζ, on K_0 and it is not hard to see that the metric space completion $K_0(\zeta)$ of K_0 with respect to d_ζ is a complete separable complex semi-inner product space. For example, if x is an element of $K_0(\zeta)$ represented by the d_ζ-Cauchy sequence (x_n) in K_0 and if a is a complex number then ax is the element of $K_0(\zeta)$ represented by the d_ζ-Cauchy sequence $(a_n x_n)$, where (a_n) is any sequence in $Q + iQ$ converging to a. Let $H(\zeta)$ be the complex Hilbert space associated with $K_0(\zeta)$ and let Λ_ζ be the natural $(Q + iQ)$-linear mapping of K_0 into $H(\zeta)$. Then the family $(\Lambda_\zeta(x))_{x \in K_0}$ is total in $H(\zeta)$ for each point ζ in Z and $\langle \Lambda_\zeta(x), \Lambda_\zeta(y) \rangle = h_{x,y}(\zeta)$ is Borel function of ζ on Z for all x, y in K_0. So (by Proposition 8.1) H is a Borel field of Hilbert spaces on Z and there is a coherence α for H such that $\zeta \mapsto \Lambda_\zeta(x)$ is an α-Borel vector field over H for each x in K_0.

From Lemmas 7.3 and 8.2 and the fact that

$$\langle T_\phi x, T_\psi y \rangle = \int_Z (\phi\bar\psi)(\zeta) h_{x,y}(\zeta) d\mu(\zeta)$$
$$= \int_Z \langle \phi(\zeta)\Lambda_\zeta(x), \psi(\zeta)\Lambda_\zeta(y) \rangle \, d\mu(\zeta)$$

for all ϕ and ψ in $L^\infty(\mu)$ and all x and y in K_0 it follows easily that there is a linear isometry W of K onto $L^2(\mu;H,\alpha)$ satisfying

$$W(T_\phi x) = \int_Z^\alpha \phi(\zeta)\Lambda_\zeta(x) d\mu(\zeta), \quad \phi \in L^\infty(\mu) \quad \text{and} \quad x \in H_0$$

It is now but a trivial calculation to verify that W is the desired isometry. □

A *direct integral decomposition* of a separable Hilbert space H with respect to an abelian von Neumann algebra M acting on H is by definition a direct integral Hilbert space whose algebra of diagonalizable operators is spatially isomorphic to M. Theorem 9.1 and the following theorem together assert that there is an essentially unique direct integral decomposition of a given separable Hilbert space H with respect to a given abelian von Neumann algebra M acting on it, and one may therefore speak of *the* direct integral decomposition of H with respect to M.

THEOREM 9.2. Let X and Y be two standard Borel spaces, let μ and ν be finite Borel measures on X and Y, respectively, let H and K be Borel fields of Hilbert spaces on X and Y, respectively, let α and β be coherences for H and K, respectively, and let M and N be the algebras of diagonalizable operators on $L^2(\mu;H,\alpha)$ and $L^2(\nu;K,\beta)$, respectively. Suppose that there is a linear isometry W of $L^2(\nu;K,\beta)$ onto $L^2(\mu;H,\alpha)$ satisfying $WNW^{-1} = M$. Then

(a) there is a Borel function ω from X to Y, a Borel function θ from Y to $(0,\infty)$ with $\omega_*(\mu) = \theta\nu$, a μ-null Borel set M in X, and a ν-null Borel set N in Y such that ω restricts to a Borel isomorphism of $X - M$ onto $Y - N$ and such that

$$W \left[\int_Y^\beta \psi(\xi) I_{K(\xi)} d\nu(\xi) \right] W^{-1} = \int_X^\alpha \psi(\omega(\zeta)) I_{H(\zeta)} d\mu(\zeta)$$

for all $\psi \in L^\infty(\nu)$, and

9. DIRECT INTEGRAL DECOMPOSITIONS 33

(b) for each point ζ in X there is a linear operator $V(\zeta)$ from $K(\omega(\zeta))$ to $H(\zeta)$ such that $V(\zeta)$ is isometric and onto $H(\zeta)$ for each $\zeta \in X - M$ and such that for any function w in $L^2(\nu;K,\beta)$ the function

$$\zeta \mapsto \theta(\omega(\zeta))^{-\frac{1}{2}} V(\zeta)w(\omega(\zeta))$$

is in $L^2(\mu;H,\zeta)$ and

$$W \int_Y^\beta w(\xi)d\nu(\xi) = \int_X^\alpha \theta(\omega(\zeta))^{-\frac{1}{2}} V(\zeta)w(\omega(\zeta))d\mu(\zeta)$$

Proof. The hypothesis and Theorem 4.6 imply that ω, θ, M, and N satisfying (a) exist.

For each function w in $L^2(\nu;K,\beta)$ choose a function v_w in $L^2(\mu;H,\alpha)$ with $W \int_Y^\beta w(\xi)d\nu(\xi) = \int_X^\alpha v_w(\zeta)d\mu(\zeta)$. Then necessarily

$$v_{au+bw} = av_u + bv_w \quad \mu\text{-a.e.}$$

for any two complex numbers a and b and any two functions u and w in $L^2(\nu;K,\beta)$. Moreover, for any function w in $L^2(\nu;K,\beta)$ one has

$$\|v_w(\zeta)\| = \theta(\omega(\zeta))^{-\frac{1}{2}} \|w(\omega(\zeta))\| \quad \text{for } \mu\text{-a.a. } \zeta \in X$$

Indeed, if ϕ is an arbitrary function in $L^\infty(\mu)$ and if ψ is a function in $L^\infty(\nu)$ with $\psi \circ \omega = \phi$ μ-a.e. then

$$\int_X |\phi(\zeta)|^2 \|v_w(\zeta)\|^2 \, d\mu(\zeta) = \left\| \int_X^\alpha \phi(\zeta) I_{H(\zeta)} d\mu(\zeta) \int_X^\alpha v_w(\zeta) d\mu(\zeta) \right\|^2$$

$$= \left\| \int_Y^\beta \psi(\xi) I_{K(\xi)} d\nu(\xi) \int_Y^\beta w(\xi) d\nu(\xi) \right\|^2$$

$$= \int_Y |\psi(\xi)|^2 \|w(\xi)\|^2 d\nu(\xi)$$

$$= \int_X \theta(\omega(\zeta))^{-1} |\phi(\zeta)|^2 \|w(\omega(\zeta))\|^2 d\mu(\zeta)$$

By Theorem 7.1 one can find a countable dense subset W of $L^2(\nu;K,\beta)$ which is a vector space over $Q + iQ$ and is such that the family $(w(\xi))_{w \in W}$ is dense in $K(\xi)$ for each ξ in Y. It follows from the above discussion that the sets M and N can be chosen so that for each ξ in $X - M$ the map $w(\omega(\zeta)) \mapsto \theta(\omega(\zeta))^{\frac{1}{2}} v_w(\zeta)$ is a well-defined $(Q + iQ)$-linear isometry of $\{w(\omega(\zeta)): w \in W\}$ into $H(\zeta)$; let $V(\zeta)$ be the extension of this map to a linear isometry of

$K(\omega(\zeta))$ into $H(\zeta)$. And for each ζ in M let $V(\zeta)$ be the zero-mapping of $K(\omega(\zeta))$ into $H(\zeta)$.

Consider a function w in $L^2(\nu;K,\beta)$. From Section 7 and the proof of Theorem 5.1 one knows that there will be a sequence (w_n) in $\mathcal{W}$ such that $\lim_{n\to\infty} w_n(\xi) = w(\xi)$ for μ-a.a. $\xi \in Y$. Thus

$$V(\zeta)w(\omega(\zeta)) = \lim_{n\to\infty} \theta(\omega(\zeta))^{\frac{1}{2}} v_{w_n}(\zeta) \quad \text{for } \mu\text{-a.a.} \quad \zeta \in X$$

and so the map

$$\zeta \mapsto \theta(\omega(\zeta))^{-\frac{1}{2}} V(\zeta)w(\omega(\zeta))$$

is in $L^2(\mu;H,\alpha)$. This shows that the map

$$\int_Y^\beta w(\xi)d\nu(\xi) \mapsto \int_X^\alpha \theta(\omega(\zeta))^{-\frac{1}{2}} V(\zeta)w(\omega(\zeta))d\mu(\zeta), \quad w \in L^2(\nu;K,\beta)$$

is a well-defined linear isometry of $L^2(\nu;K,\beta)$ into $L^2(\mu;H,\alpha)$ which agrees with W on a dense set.

It remains to show that $V(\zeta)$ is onto $H(\zeta)$ for μ-a.a. $\zeta \in X$. Now for this it is enough (by Theorem 7.1) to show that $\zeta \mapsto V(\zeta)V(\zeta)^*$ is an α-Borel operator field and that $\int_X^\alpha V(\zeta)V(\zeta)^*d\mu(\zeta)$ is the identity operator. If v is any α-Borel vector field over H and if x is any vector in ℓ^2 then

$$\langle \alpha(\zeta)V(\zeta)V(\zeta)^*v(\zeta),x\rangle = \sum_j \langle v(\zeta),V(\zeta)\beta(\omega(\zeta))^*e_j\rangle \langle \alpha(\zeta)V(\zeta)\beta(\omega(\zeta))^*e_j,x\rangle$$

(here (e_j) is as in Section 8) is a Borel function of ζ on Z as $\beta(\cdot)^*e_j \in L^2(\nu;K,\beta)$. This shows that $\zeta \mapsto V(\zeta)V(\zeta)^*$ is α-Borel, and by construction $\int_X^\alpha V(\zeta)V(\zeta)^*d\mu(\zeta)$ agrees with the identity operator on a dense set. □

10. Examples

This section contains a number of examples of direct integral decompositions of separable Hilbert spaces with respect to abelian von Neumann algebras. Such decompositions can, at least in principle, be deduced from the known structure of von Neumann algebras of type I (Theorem A.6). As will become clear from the following examples, however, this is not a useful technique for obtaining direct integral decompositions; it does, nevertheless, shed some light on such decompositions and leads to a new proof of Theorem 9.1.

10. EXAMPLES

EXAMPLE 10.1. The direct integral decomposition of a separable Hilbert space H with respect to the trivial algebra $C(H)$ is just $L^2(\mu;H)$, where μ is a non-zero finite measure on some one-point space. □

EXAMPLE 10.2. Suppose that M is a maximal abelian von Neumann algebra acting on a separable Hilbert space H and that M is *-isomorphic to $L^\infty(\mu)$ for some Borel measure μ on a standard Borel space Z. If $\phi \mapsto T_\phi$ is the *-homomorphism of $L^\infty(\mu)$ onto M obtained by composing the canonical mapping of $L^\infty(\mu)$ onto $L^\infty(\mu)$ with some fixed *-isomorphism of $L^\infty(\mu)$ onto M (cf. Theorem A.4) and if x is a cyclic and separating vector for M (such a vector exists by a standard exhaustion argument) then there is a Borel function θ from Z to $(0,\infty)$ with $<T_\phi x, x> = \int_Z \phi\theta\, d\mu$ for all $\phi \in L^\infty(\mu)$ (cf. the proof of Theorem 9.1). Thus $\theta \in L^1(\mu)$ and $L^\infty(\mu) = L^\infty(\theta\mu) \subset L^2(\theta\mu)$, and there is a linear isometry W of $L^2(\theta\mu)$ onto H satisfying $W\phi = T_\phi x$, $\phi \in L^\infty(\theta\mu)$. If W also denotes the corresponding linear isometry of $L^2(\theta\mu)$ onto H then $W^{-1}T_\phi W$ is, for each $\phi \in L^\infty(\theta\mu)$, the operator on $L^2(\theta\mu)$ determined by pointwise multiplication by ϕ on $L^2(\theta\mu)$. This means that $L^2(\theta\mu)$ and (by Lemma 7.2) $L^2(\mu)$ are both direct integral decomposition of H with respect to M. □

EXAMPLE 10.3. Now suppose that M is an arbitrary abelian von Neumann algebra acting on a separable Hilbert space H. Then there will be a subset J of $\mathbb{N} \cup \{\infty\}$ and a family $(M_j)_{j \in J}$ of abelian von Neumann algebras such that M' and $\oplus_{j \in J} [M_j \otimes L(\ell_j^2)]$ are spatially isomorphic (Theorems A.6 and A.9). But then M and $\oplus_{j \in J} [M_j' \otimes C(\ell_j^2)]$ too will be spatially isomorphic, and so each M_j' must be abelian. This means that each M_j satisfies $M_j = M_j'$, i.e., is a maximal abelian von Neumann algebra on the Hilbert space on which it acts, say H_j. So from Theorem A.4 and the preceeding example there is a standard Borel space Z_j and a finite Borel measure μ_j on Z_j such that $L^2(\mu_j)$ is the direct integral decomposition of H_j with respect to M_j. But then it follows from Proposition 5.2 that $L^2(\mu_j;\ell_j^2)$ is the direct integral decomposition of $H_j \otimes \ell_j^2$ with respect to $M_j \otimes C(\ell_j^2)$. Now let Z be the disjoint union of the Z_j with the sum Borel structure, let μ be the sum of the μ_j (i.e., $\mu(A) = \sum_j \mu_j(A \cap Z_j)$ for each Borel set A in Z), let $H(\zeta) = \ell_j^2$ for $\zeta \in Z_j$, and let $\alpha(\zeta)$ be the identity operator on $H(\zeta)$ for each $\zeta \in Z$. Then it is easy to see that there is a linear isometry of $L^2(\mu;H,\alpha)$ onto $\oplus_{j \in J} [H_j \otimes \ell_j^2]$ which carries the algebra of diagonalizable operators onto $\oplus_{j \in J} [M_j \otimes C(\ell_j^2)]$. This means, finally, that $L^2(\mu;H,\alpha)$ is the direct integral decomposition of H with respect to M. □

The next five examples illustrate how one can use commutative harmonic analysis to find direct integral decompositions in the case where the Hilbert space in question is the L^2-space of a locally compact group and the abelian von Neumann algebra in question arises from the regular representation of the group. The first two of these examples are included for the benefit of the reader who is unfamiliar with commutative harmonic analysis (i.e., with the Haar measure, the Fourier transform, and the Plancherel theorem for a locally compact abelian group); the reader who is familiar with this general theory should proceed directly to Example 10.6.

EXAMPLE 10.4. Let μ be normalized Lebesgue measure on the circle group T. It is known from the theory of Fourier series that the formula

$$(Wf)(n) = \frac{1}{2\pi} \int_0^{2\pi} f(e^{i\theta}) e^{-in\theta} \, d\theta, \quad f \in L^2(\mu) \text{ and } n \in \mathbb{Z}$$

defines a linear isometry W of $L^2(\mu)$ onto $\ell^2(\mathbb{Z})$. If λ denotes the left regular representation of T (see Appendix C) then an easy calculation will show that

$$[W\lambda(e^{ir})W^{-1}g](n) = e^{-inr} g(n)$$

for all $r \in [0, 2\pi]$, $n \in \mathbb{Z}$, and $g \in \ell^2(\mathbb{Z})$. As the functions $n \mapsto e^{inr}$, $r \in [0, 2\pi]$, generate $\ell^\infty(\mathbb{Z})$ as a von Neumann algebra, one concludes that $\ell^2(\mathbb{Z})$ is the direct integral decomposition of $L^2(\mu)$ with respect to $\lambda(T)''$. □

EXAMPLE 10.5. Let μ be Lebesgue measure on the real line $\mathbb{R}$. From the theory of the Fourier transform it is known that the formula

$$(Wf)(r) = \int_{-\infty}^{\infty} f(s) e^{-irs} \, ds, \quad f \in L^1(\mu) \cap L^2(\mu) \text{ and } r \in \mathbb{R}$$

defines a linear isometry W of $L^2(\mu)$ onto $L^2(\frac{1}{2\pi}\mu)$. If λ denotes the left regular representation of $\mathbb{R}$ (see Appendix C) then $W\lambda(r)W^{-1}$ is, for each $r \in \mathbb{R}$, the operator on $L^2(\frac{1}{2\pi}\mu)$ determined by pointwise multiplication by $s \mapsto e^{-irs}$ on $L^2(\frac{1}{2\pi}\mu)$. As the images of the functions $s \mapsto e^{irs}$, $r \in \mathbb{R}$, in $L^\infty(\mu)$ generate $L^\infty(\mu)$ as a von Neumann algebra, one concludes that $L^2(\frac{1}{2\pi}\mu)$ is the direct integral decomposition of $L^2(\mu)$ with respect to $\lambda(\mathbb{R})''$. □

EXAMPLE 10.6. This example includes the two preceeding ones as special cases. Let μ be a Haar measure on a separable locally compact abelian group G, let

10. EXAMPLES

$\hat{G}$ be the dual group, and let ν be a Haar measure on $\hat{G}$. According to the Plancherel theorem, ν can be chosen so that there is a linear isometry W of $L^2(\mu)$ onto $L^2(\nu)$ whose restriction to $L^2(\mu) \cap L^1(\mu)$ is induced by the Fourier transform, i.e.,

$$(Wf)(\gamma) = \int_G f(r)\overline{\gamma(r)}d\mu(r), \quad f \in L^1(\mu) \cap L^2(\mu) \text{ and } \gamma \in \hat{G}$$

If λ denotes the left regular representation of G (see Appendix C) then $W\lambda(r)W^{-1}$ is, for each $r \in G$, the operator on $L^2(\nu)$ determined by pointwise multiplication by $\gamma \mapsto \overline{\gamma(r)}$ on $L^2(\nu)$. Now the images of the functions $\gamma \mapsto \gamma(r)$, $r \in G$, in $L^\infty(\nu)$ generate $L^\infty(\nu)$ as a von Neumann algebra, and hence $L^2(\nu)$ is the direct integral decomposition of $L^2(\mu)$ with respect to $\lambda(G)''$. □

The next two examples may appear at first sight to be somewhat contrived. In actual fact, they are related to Mackey's theory of induced representations and should even be familiar to anyone who is acquainted with this theory.

EXAMPLE 10.7. In this example K will denote a countably infinite discrete field (e.g., the rationals, or the algebraic completion of a finite field) and K* the multiplicative group of nonzero elements of K. It is easy to verify that the set $G = K^* \times K$ becomes a group if one defines multiplication by

$$(a,b)(c,d) = (ac, ad+b)$$

Let $(\delta_r)_{r \in G}$, $(\varepsilon_a)_{a \in K^*}$, and $(\eta_b)_{b \in K}$ be the canonical orthonormal basis in $\ell^2(G)$, $\ell^2(K^*)$, and $\ell^2(K)$, respectively, let λ and λ' be the left and right regular representations of G, respectively (see Appendix C), and let μ and ν be the normalized Haar measures on the compact groups $X = (K^*)^{\hat{}}$ and $Y = \hat{K}$, respectively, where K is regarded as a group under addition. As the functions $\zeta \mapsto \zeta(a)$, $a \in K^*$, form an orthonormal basis in $L^2(\mu)$, Proposition 5.2 implies that there is a linear isometry V of $\ell^2(G)$ onto $L^2(\mu; \ell^2(K))$ satisfying

$$V\delta_{(a,b)} = \int_X^\oplus \zeta(a)\eta_b d\mu(\zeta), \quad (a,b) \in G$$

It is a simple matter to verify that

$$V\lambda'(c,0)V^{-1} = \int_X^\oplus \zeta(c^{-1})I \, d\mu(\zeta), \quad c \in K^*$$

As the images of the functions $\zeta \mapsto \zeta(c)$, $c \in K^*$, in $L^\infty(\mu)$ generate $L^\infty(\mu)$ as a von Neumann algebra, one sees that $L^2(\mu;\ell^2(K))$ is the direct integral decomposition of $\ell^2(G)$ with respect to $\lambda'(K^* \times \{0\})''$. A similar argument will show that there is a linear isometry W of $\ell^2(G)$ onto $L^2(\nu;\ell^2(K^*))$ satisfying

$$W\delta_{(a,b)} = \int_Y^\oplus \xi(a^{-1}b)\varepsilon_a \, d\nu(\xi), \quad (a,b) \in G$$

that

$$W\lambda'(1,d)W^{-1} = \int_Y^\oplus \xi(-d)I \, d\nu(\xi), \quad d \in K$$

and that $L^2(\nu;\ell^2(K^*))$ is the direct integral decomposition of $\ell^2(G)$ with respect to $\lambda'(\{1\} \times K)''$.

Given an element (a,b) in G, the operator $\lambda(a,b)$ lies in $\lambda'(G)'$ and so both $V\lambda(a,b)V^{-1}$ and $W\lambda(a,b)W^{-1}$ are decomposable by Theorem 6.2. In fact, it is not difficult to verify that

$$V\lambda(a,b)V^{-1} = \int_X^\oplus J_{(a,b)}(\zeta)d\mu(\zeta)$$

where

$$J_{(a,b)}(\zeta)\eta_d = \zeta(a)\eta_{ad+b}, \quad \zeta \in X \text{ and } d \in K$$

and that

$$W\lambda(a,b)W^{-1} = \int_Y^\oplus K_{(a,b)}(\xi)d\nu(\xi)$$

where

$$K_{(a,b)}(\xi)\varepsilon_c = \xi(a^{-1}c^{-1}b)\varepsilon_{ac}, \quad \xi \in Y \text{ and } c \in K^* \quad \square$$

EXAMPLE 10.8. This example contains the preceeding one as a special case. Let H be an abelian subgroup of a countably infinite discrete group G and let μ be the normalized Haar measure on the compact group $Z = \hat{H}$. Let K be any subset of G which meets each left H-coset in exactly one point. For each element ζ of Z let $H(\zeta)$ be the set consisting of all those complex-valued functions f on G satisfying

(a) $f(rp) = \zeta(p^{-1})f(r)$ for all $r \in G$ and $p \in H$, and
(b) $\sum_{s \in K} |f(s)|^2 < \infty$

notice that the sum in (b) is independent of the particular choice of K and that $H(\zeta)$ is actually a Hilbert space which is isomorphic to $\ell^2(K)$ in a natural way. Let $(\delta_r)_{r \in G}$ be the canonical orthonormal basis in $\ell^2(G)$ and

10. EXAMPLES

for each r in G define a vector field $W\delta_r$ over H by putting

$$(W\delta_r)(\zeta)(s) = \begin{cases} 0 & s^{-1}r \notin H \\ \zeta(s^{-1}r) & s^{-1}r \in H \end{cases}$$

for all $\zeta \in \hat{H}$ and $s \in G$. Then $((W\delta_r)(\zeta))_{r \in K}$ is an orthonormal basis in $H(\zeta)$ and $\zeta \mapsto <(W\delta_r)(\zeta),(W\delta_s)(\zeta)>$ is a Borel function on Z whose integral with respect to μ is (δ_r,δ_s) for all choices of r and s in G. There is therefore (by Proposition 8.1) a coherence α for H such that each $W\delta_r$ is α-Borel, and as $L^2(\mu;H,\alpha)$ is complete (see Theorem 7.1) W extends to a linear isometry, which will still be denoted by W, of $\ell^2(G)$ into $L^2(\mu;H,\alpha)$. Actually, W is onto a dense subset of $L^2(\mu;H,\alpha)$. To see this, take an element v in $L^2(\mu;H,\alpha)$ which is orthogonal to each $W\delta_r$. Then

$$0 = <v,W\delta_{rp}> = \int_Z <v(\zeta),(W\delta_{rp})(\zeta)> d\mu(\zeta)$$

$$= \int_Z v(\zeta)(r)\zeta(p^{-1}) d\mu(\zeta)$$

for all r in K and all p in H, and so $v(\cdot)(r) = 0$ μ-a.e. for each fixed $r \in K$. It now follows immediately from the countability of G that $v = 0$ μ-a.e.

Let λ and λ' denote the left and right regular representations of G, respectively (see Appendix C). An easy calculation will show that

$$W\lambda'(p)W^{-1} = \int_Z^\alpha \zeta(p^{-1}) I_{H(\zeta)} d\mu(\zeta), \quad p \in H$$

as the images of the functions $\zeta \mapsto \zeta(p)$, $p \in H$, in $L^\infty(\mu)$ generate $L^\infty(\mu)$ as a von Neumann algebra, this calculation shows that $L^2(\mu;H,\alpha)$ is the direct integral decomposition of $\ell^2(G)$ with respect to $\lambda'(H)''$. Now given an element $r \in G$, the operator $\lambda(r)$ lies in $\lambda'(H)'$ and hence $W\lambda(r)W^{-1}$ is decomposable by Theorem 7.1. In fact, it is not hard to verify that

$$W\lambda(r)W^{-1} = \int_Z^\alpha J_r(\zeta) d\mu(\zeta)$$

where

$$[J_r(\zeta)f](s) = f(r^{-1}s), \quad f \in H(\zeta), \quad s \in G, \text{ and } \zeta \in Z \quad \square$$

The last example is concerned with the very natural question of how the direct integral decompositions of a separable Hilbert space with respect to two abelian von Neumann algebras acting on it are related if one of the algebras in question is a subalgebra of the other. Roughly speaking, the answer is that the larger the abelian algebra in question is the larger the base space and the finer the component Hilbert spaces will be (cf. the first two examples); putting it a bit more precisely, if H_0 is the Hilbert space and M and N the abelian von Neumann algebras in question and if $N \subset M$, then each of the Hilbert spaces appearing in the decomposition of H_0 with respect to N can itself be decomposed with respect to some abelian von Neumann algebra acting on it so as to obtain all of the Hilbert spaces appearing in the decomposition of H_0 with respect to M. This will be made quite precise in a moment and will be further elucidated in Example 20.2. In principle, the result of this example combined with Example 10.2 would enable one to construct the direct integral decomposition of any separable Hilbert space with respect to any abelian von Neumann algebra acting on it; in fact, one could give an alternative proof of Theorem 9.1 along these lines, a proof which would be less instructive than the original one.

EXAMPLE 10.9. Let H_0, M, and N be as in the preceeding paragraph. Then by Theorem 9.1 there will be a standard Borel space X [respectively, Y], a finite Borel measure μ on X [respectively, ν on Y], a Borel field of Hilbert spaces H on X [respectively, K on Y], a coherence α for H [respectively, β for K], and a linear isomorphism U [respectively, V] of H_0 onto $L^2(\mu;H,\alpha)$ [respectively, $L^2(\nu;K,\beta)$] such that UMU^{-1} [respectively, VNV^{-1}] is the algebra of diagonalizable operators on $L^2(\mu;H,\alpha)$ [respectively, $L^2(\mu;K,\beta)$] and such that

$$\phi \mapsto U^{-1} \int_X^\alpha \phi(\zeta) I_{H(\zeta)} d\mu(\zeta)\, U$$

[respectively,

$$\psi \mapsto V^{-1} \int_Y^\beta \psi(\xi) I_{K(\xi)} d\nu(\xi)\, V]$$

is the *-homomorphism of $L^\infty(\mu)$ onto M [respectively, of $L^\infty(\nu)$ onto N] which is obtained by composing the canonical mapping of $L^\infty(\mu)$ onto $L^\infty(\mu)$ [respectively, of $L^\infty(\nu)$ onto $L^\infty(\nu)$] with some fixed *-isomorphism of $L^\infty(\mu)$ onto M [respectively, of $L^\infty(\nu)$ onto N]. Next, one can apply Theorem 4.6 to the inclusion mapping of N into M to obtain (after possibly deleting a μ-null Borel set from X, a ν-null Borel set from Y, and replacing ν by an equivalent measure (the last change of which is justified by Lemma 7.2)) a Borel map ω of X onto Y satisfying $\omega_*(\mu) = \nu$ and

10. EXAMPLES

$$UV^{-1}\left[\int_Y^\beta \psi(\xi) I_{K(\xi)} d\nu(\xi)\right] VU^{-1} = \int_X^\alpha \psi(\omega(\zeta)) I_{H(\zeta)} d\mu(\zeta)$$

for all $\psi \in L^\infty(\nu)$, and then Theorem 4.5 to X, Y, μ, and ω to obtain a Borel map $\xi \mapsto \mu_\xi$ from Y to M(X) satisfying the two conditions of that theorem.

From Proposition 5.2 and the discussion in Section 7 it is clear that one can find a sequence (v_m) in $L^2(\mu; H, \alpha)$ such that $\|v_m(\cdot)\|$ is a bounded function on X for each m and such that the sequence $(v_m(\zeta))$ is dense in $H(\zeta)$ for each $\zeta \in X$. Now as X is a standard Borel space there is a compact metric topology on X generating the given Borel structure (use Theorem 2.14), and hence there will be a sequence (ϕ_j) of Borel functions from X to the closed unit disc in the complex plane whose image in $L^\infty(\lambda)$ is weak *-dense in the unit ball of $L^\infty(\lambda)$ for each finite Borel measure λ on X. One can now invoke Lemma 7.3 to conclude that the vectors $\int_X^\alpha \phi_j(\zeta) v_m(\zeta) d\mu_\xi(\zeta)$ are total in $L^2(\mu_\xi; H, \alpha)$ for each $\xi \in Y$. Moreover,

$$\langle \int_X^\alpha \phi_j(\zeta) v_m(\zeta) d\mu_\xi(\zeta), \int_X^\alpha \phi_k(\zeta) v_n(\zeta) d\mu_\xi(\zeta) \rangle$$
$$= \int_X (\phi_j \overline{\phi_k})(\zeta) \langle v_m(\zeta), v_n(\zeta) \rangle d\mu_\xi(\zeta)$$

is a Borel function of ξ on Y for all choices of j, k, m and n, and so, by Proposition 8.1, $\xi \mapsto L^2(\mu_\xi; H, \alpha)$ is a Borel field of Hilbert spaces on Y and there is a coherence γ for this field such that $\xi \mapsto \int_X^\alpha \phi_j(\zeta) v_m(\zeta) d\mu_\xi(\zeta)$ is a γ-Borel vector field for all j and m. Now consider an α-Borel vector field v over H with the property that $\|v(\cdot)\|$ is a bounded function (such v are dense in $L^2(\mu; H, \alpha)$ as μ is finite). Then v belongs to $L^2(\mu_\xi; H, \alpha)$, $\xi \in Y$, as each μ_ξ is finite, and

$$\langle \int_X^\alpha v(\zeta) d\mu_\xi(\zeta), \int_X^\alpha \phi_j(\zeta) v_m(\zeta) d\mu_\xi(\zeta) \rangle$$
$$= \int_X \overline{\phi_j}(\zeta) \langle v(\zeta), v_m(\zeta) \rangle d\mu_\xi(\zeta)$$

is a Borel function of ξ on Y for all j and m. Thus $\xi \mapsto \int_X^\alpha v(\zeta) d\mu_\xi(\zeta)$ is a γ-Borel vector field by Lemma 8.2, and in fact belongs to $L^2(\nu; \xi \mapsto L^2(\mu_\xi; H, \alpha), \gamma)$ as

$$\int_Y \|\int_X^\alpha v(\zeta) d\mu_\xi(\zeta)\|^2 d\nu(\xi) = \int_X \|v(\zeta)\|^2 d\mu(\zeta) < \infty$$

by part (b) of Theorem 4.5. In view of Lemma 7.3, it should now be clear that there is a well-defined linear isometry W of $\int_X^\alpha H(\zeta) d\mu(\zeta)$ onto

$\int_Y^\gamma \int_X^\alpha H(\zeta) d\mu_\xi(\zeta) d\nu(\xi)$ satisfying

$$W \int_X^\alpha v(\zeta) d\mu(\zeta) = \int_Y^\gamma \int_X^\alpha v(\zeta) d\mu_\xi(\zeta) d\nu(\xi)$$

for all those functions v in $L^2(\mu; H, \alpha)$ for which $\|v(\cdot)\|$ is bounded. An easy calculation using part (a) of Theorem 4.5 now shows that

$$W \left[\int_X^\alpha \psi(\omega(\zeta)) I_{H(\zeta)} d\mu(\zeta) \right] W^{-1} = \int_Y^\gamma \psi(\xi) I \, d\nu(\xi)$$

for all $\psi \in L^\infty(\nu)$. From this one can conclude, finally, that $\int_Y^\gamma \int_X^\alpha H(\zeta) d\mu_\xi(\zeta) d\nu(\xi)$ is a direct integral decomposition of $L^2(\mu; H, \alpha)$ with respect to UNU^{-1}.

This means that $\int_Y^\beta K(\xi) d\nu(\xi)$ and $\int_Y^\gamma \int_X^\alpha H(\zeta) d\mu_\xi(\zeta) d\nu(\xi)$ are both direct integral decompositions of H with respect to N, and so one can invoke the uniqueness theorem (Theorem 9.2). But first notice that

$$(WUV^{-1}) \left[\int_Y^\beta \psi(\xi) I_{K(\xi)} d\nu(\xi) \right] (WUV^{-1})^{-1}$$

$$= W \left[\int_X^\alpha \psi(\omega(\zeta)) I_{H(\zeta)} d\mu(\zeta) \right] W^{-1}$$

$$= \int_Y^\gamma \psi(\xi) \mathrm{Id}\nu(\xi)$$

for any $\psi \in L^\infty(\nu)$. So by Theorem 9.2 there is (after possibly deleting yet another ν-null Borel set from Y) for each $\xi \in Y$ a linear isometry $T(\xi)$ of $K(\xi)$ onto $L^2(\mu_\xi; H, \alpha)$ such that the coherences β and $\xi \mapsto \gamma(\xi) T(\xi)$ are equivalent and such that

$$(WUV^{-1}) \int_Y^\beta w(\xi) d\nu(\xi) = \int_Y^\gamma T(\xi) w(\xi) d\nu(\xi)$$

for all $w \in L^2(\nu; K, \beta)$. This means that from the point of view of direct integral theory, one may as well identify the fields K and $\xi \mapsto L^2(\mu_\xi; H, \alpha)$ and the coherences β and γ (cf. the discussion in Section 8). □

HISTORICAL COMMENTS

The theory of direct integrals of Hilbert spaces and operators was first studied by von Neumann in the 1930s although his results were not published until 1949 [49]. More recently, they have been discussed in a number of papers and books: in particular, the books of Dixmier [10], Naimark [47],

10. EXAMPLES

and Schwartz [54] contain fairly complete and systematic expositions of the theory. There are, however, only superficial differences between the various accounts (including this one) of the theory.

Although coherences are certainly implicit in all treatments of direct integral theory, it was Effros who (in [16]) brought them to the forefront and made them part of the notation. For example, Section 8 is essentially taken from [10]. Whether coherences are implicit or explicit does not affect how one thinks about or what one can do with direct integrals, yet making them explicit does remove some ambiguity from the notation and makes some statements a bit more precise.

Most of the results of this chapter save those in Sections 8 and 10 can be found in von Neumann's paper on the theory of direct integrals [49]. In particular, all of Theorem 7.1 and both Theorems 9.1 and 9.2 appear therein. The examples presented in Section 10 are well-known. Example 10.7 was first mentioned by Mackey [38, pp. 590-591], and Example 10.9 is due to Mackey [39, Theorem 2.11] and can also be found in [16, Lemma 4.5; 22, Section 5; 50, pp. 600-602].

Chapter 3
DIRECT INTEGRALS OF REPRESENTATIONS

11. Definitions And Some Elementary Properties

Throughout this section R will denote a separable involutive Banach algebra and G a separable locally compact group. The reader should consult Appendices B and C for the definitions and properties of representations of R and G.

Consider a Borel space Z, a Borel measure μ on Z, and a Borel field of Hilbert spaces H on Z. A *field of representations of* R [respectively, G] *over* H is by definition a function π assigning to each point ζ in Z a representation $\pi(\zeta)$ of R [respectively, G] on $H(\zeta)$. If, in addition, α is a coherence for H then such a field of representations π will be called α-*Borel* if each $\pi(\cdot)(R)$, $R \in R$, [respectively, each $\pi(\cdot)(r)$, $r \in G$] is an α-Borel operator field over H. Assuming that π is an α-Borel field of representations of R [respectively, G] over H, it follows from Theorem 7.1 and Proposition B.1 that there is a representation of R [respectively, G] on $L^2(\mu;H,\alpha)$ whose value at an R in R [respectively, at an r in G] is $\int_Z^\alpha \pi(\zeta)(R)d\mu(\zeta)$ [respectively, $\int_Z^\alpha \pi(\zeta)(r)d\mu(\zeta)$]. This representation is called the *direct integral of* π *with respect to* μ *and* α and will be denoted by $\int_Z^\alpha \pi(\zeta)d\mu(\zeta)$ or, in case both H and α are constant, by $\int_Z^\oplus \pi(\zeta)d\mu(\zeta)$. Notice that the algebra of diagonalizable operators on $L^2(\mu;H,\alpha)$ is contained in the commutant of the range of $\int_Z^\alpha \pi(\zeta)d\mu(\zeta)$. The representation $\int_Z^\alpha \pi(\zeta)d\mu(\zeta)$ will also be called a *direct integral decomposition* or a representation ρ of R [respectively, G] on a Hilbert space K with respect to an abelian von Neumann algebra M contained in $\rho(R)'$ [respectively, $\rho(G)'$] if there is a linear isometry W of K onto $L^2(\mu;H,\alpha)$ with $W\rho(\cdot)W^{-1} =$

$\int_Z^\alpha \pi(\zeta)d\mu(\zeta)$ and such that WMW^{-1} is the algebra of diagonalizable operators on $L^2(\mu;H,\alpha)$; and moreover, the decomposition will be called *maximal* if M is a maximal abelian subalgebra of $\rho(R)'$ [respectively, $\rho(G)'$] and *central* if M is the center of $\rho(R)'$ [respectively, $\rho(G)'$]. Letting μ_n denote the restriction of μ to the Borel subsets of the Borel set $Z_n = \{\zeta \in Z: \dim H(\zeta) = n\}$, $n = \infty, 0, 1, 2, \ldots$, and letting W be the linear isometry of $L^2(\mu;H,\alpha)$ onto $\oplus_{0 \le n \le \infty} L^2(\mu_n; \ell_n^2)$ constructed in Section 7, it is not difficult to verify that

$$W\left[\int_Z^\alpha \pi(\zeta)d\mu(\zeta)\right](\cdot)W^{-1} = \oplus_{0 \le n \le \infty} \int_{Z_n}^\oplus \pi_n(\zeta)d\mu_n(\zeta) \qquad (*)$$

where $\pi_n(\zeta)(\cdot) = \alpha(\zeta)\pi(\zeta)(\cdot)\alpha(\zeta)^*|\ell_n^2$ for each $\zeta \in Z_n$. Finally, notice that if Z is discrete and if μ is counting measure on Z then $\int_Z^\alpha \pi(\zeta)d\mu(\zeta)$ is nothing but $\oplus_{\zeta \in Z} \pi(\zeta)$.

PROPOSITION 11.1. Suppose that Z is a Borel space, that μ is a Borel measure on Z, that H is a Borel field of Hilbert spaces on Z, that α is a coherence for H, and that π is an α-Borel field of representations of R over H. If $\pi(\zeta)$ is nondegenerate for μ-a.a. $\zeta \in Z$ then $\int_Z^\alpha \pi(\zeta)d\mu(\zeta)$ is nondegenerate and, conversely, if R has a sequential approximate identity and if $\int_Z^\alpha \pi(\zeta)d\mu(\zeta)$ is nondegenerate then $\pi(\zeta)$ is nondegenerate for μ-a.a. $\zeta \in Z$.

Proof. First suppose that $\pi(\zeta)$ is nondegenerate for μ-a.a. $\zeta \in Z$, and let v be a function in $L^2(\mu;H,\alpha)$ such that $\int_Z^\alpha v(\zeta)d\mu(\zeta)$ is in the kernel of $\int_Z^\alpha \pi(\zeta)(R)d\mu(\zeta)$ for each $R \in R$. Then for each $R \in R$ there is a μ-null Borel set N_R in Z with $\pi(\zeta)(R)v(\zeta) = 0$ for all $\zeta \in Z - N_R$, and if $N = \cup_R N_R$, where the union is over some countable dense subset of R, then N is μ-null and Borel and, by Proposition B.1, has the property that $\pi(\zeta)(R)v(\zeta) = 0$ for all $R \in R$ and all $\zeta \in Z - N$. This shows that $v(\zeta) = 0$ for μ-a.a. $\zeta \in Z$ and proves that $\int_Z^\alpha \pi(\zeta)d\mu(\zeta)$ is nondegenerate. Conversely, if R has a sequential approximate identity and if $\int_Z^\alpha \pi(\zeta)d\mu(\zeta)$ is nondegenerate then it follows easily from Theorem 7.1(vi) and Proposition B.5 that $\pi(\zeta)$ must be nondegenerate for μ-a.a. $\zeta \in Z$. □

The next result will be very useful because it allows one to reduce many questions concerning direct integrals of representations of groups to ones concerning direct integrals of representations of algebras.

PROPOSITION 11.2. Let Z, μ, H, and α be as in Proposition 11.1. A field of representations π of G over H is α-Borel if and only if the associated field of representations $\pi(\cdot)$ of $L^1(G)$ over H is α-Borel, and in this case

$$\left[\int_Z^\alpha \pi(\zeta) d\mu(\zeta) \right]^\sim = \int_Z^\alpha \widetilde{\pi(\zeta)} d\mu(\zeta)$$

Proof. Let v and w be two α-Borel vector fields over H and let (ϕ_n) be a sequential approximate identity for $L^1(G)$ (such a (ϕ_n) exists by Proposition C.1). Then for each $r \in G$, each $f \in L^1(G)$, and each $\zeta \in Z$ one has

$$<\widetilde{\pi(\zeta)}(f)v(\zeta),w(\zeta)> = \int_G f(r)<\pi(\zeta)(r)v(\zeta),w(\zeta)>dr$$

by definition, and

$$<\pi(\zeta)(r)v(\zeta),w(\zeta)> = \lim_{n\to\infty} <\widetilde{\pi(\zeta)}(\lambda(r)\phi_n)v(\zeta),w(\zeta)>$$

by formula (2) of Appendix C, the nondegeneracy of $\widetilde{\pi(\zeta)}$, and Proposition B.5. It is clear from these formulae and Fubini's theorem that π is α-Borel if and only if $\widetilde{\pi(\cdot)}$ is. If this is the case and if v and w lie in $L^2(\mu;H,\alpha)$ then

$$<\left[\int_Z^\alpha \widetilde{\pi(\zeta)} d\mu(\zeta) \right](f) \int_Z^\alpha v(\zeta)d\mu(\zeta), \int_Z^\alpha w(\zeta)d\mu(\zeta)>$$

$$= \int_Z <\widetilde{\pi(\zeta)}(f)v(\zeta),w(\zeta)>d\mu(\zeta)$$

$$= \int_Z \int_G f(r)<\pi(\zeta)(r)v(\zeta),w(\zeta)>drd\mu(\zeta)$$

$$= \int_G f(r)<\left[\int_Z^\alpha \pi(\zeta)d\mu(\zeta) \right](r) \int_Z^\alpha v(\zeta)d\mu(\zeta), \int_Z^\alpha w(\zeta)d\mu(\zeta)>dr$$

$$= <\left[\int_Z^\alpha \pi(\zeta)d\mu(\zeta) \right]^\sim (f) \int_Z^\alpha v(\zeta)d\mu(\zeta), \int_Z^\alpha w(\zeta)d\mu(\zeta)>$$

and thus the equation in question must be true. □

12. Equivalence, Existence and Uniqueness of Direct Integral Decompositions

Let R and G continue to denote a separable involutive Banach algebra and a separable locally compact group, respectively. The results contained in this section can (roughly speaking) be summed up as follows: the direct integral can be regarded as a function on either the equivalence classes or else the quasi-equivalence classes of representations of R and G, and there is an essentially unique (and thus one may speak of *the*) direct integral decomposition of a given representation ρ of R [respectively, G] with respect to a given abelian von Neumann algebra contained in ρ(R)' [respectively, ρ(G)'].

THEOREM 12.1. Suppose that Z is a standard Borel space, that μ is a Borel measure on Z, that H and K are two Borel fields of Hilbert spaces on Z, that α and β are coherences for H and K, respectively, and that π and ρ are α-Borel and β-Borel fields of representations of either R or G over H and K, respectively. Then $\int_Z^\alpha \pi(\zeta)d\mu(\zeta)$ and $\int_Z^\beta \rho(\zeta)d\mu(\zeta)$ are equivalent if $\pi(\zeta) \simeq \rho(\zeta)$ for μ-a.a. $\zeta \in Z$ and quasi-equivalent if $\pi(\zeta) \approx \rho(\zeta)$ for μ-a.a. $\zeta \in Z$.

Proof. By Theorem C.2 and Propositions 11.1, 11.2, and C.1, it is enough to consider the case of R. Considering first the assertion concerning equivalence, one may as well assume that $\pi(\zeta) \simeq \rho(\zeta)$ for each $\zeta \in Z$, i.e., that for each point ζ in Z there is a linear isometry $U(\zeta)$ of $H(\zeta)$ onto $K(\zeta)$ intertwining $\pi(\zeta)$ and $\rho(\zeta)$. If the $U(\zeta)$ are such that $\zeta \mapsto U(\zeta)v(\zeta)$ is a β-Borel vector field over K whenever v is an α-Borel vector field over H then the formula

$$W \int_Z^\alpha v(\zeta)d\mu(\zeta) = \int_Z^\beta U(\zeta)v(\zeta)d\mu(\zeta), \quad v \in L^2(\mu;H,\alpha)$$

would define a linear isometry W of $L^2(\mu;H,\alpha)$ onto $L^2(\mu;K,\beta)$ which intertwines $\int_Z^\alpha \pi(\zeta)d\mu(\zeta)$ and $\int_Z^\beta \rho(\zeta)d\mu(\zeta)$. The proof, then, consists in showing that the $U(\zeta)$ can be chosen in this way.

As $H(\zeta)$ and $K(\zeta)$ have the same dimension for each point ζ in Z, one may (by formula (*) of Section 11) assume that H,K,α, and β are all constant and that H = K and α = β. This means that there is a Hilbert space H such that $\pi(\zeta)$ and $\rho(\zeta)$ are representations of R on H for each $\zeta \in Z$ and such that the functions $\pi(\cdot)(R)$ and $\rho(\cdot)(R)$ from Z to $L(H)$ are Borel for each $R \in R$. Let U denote the group of unitary operators on H and let

$$W = \{(\zeta,U) \in Z \times U: \pi(\zeta)(\cdot) = U\rho(\zeta)(\cdot)U^*\}$$

If (R_k) is a dense sequence in R then

12. EQUIVALENCE, EXISTENCE AND UNIQUENESS 49

$$W = \cap_k \{(\zeta,U) \in Z \times L(H): U^*U = UU^* = I \text{ and } \pi(\zeta)(R_k)U - U\rho(\zeta)(R_k) = 0\}$$

by Proposition B.1, and thus W is a Borel, and hence a standard, subset of $Z \times L(H)$. The image of W under the projection of $Z \times U$ onto its first coordinate is all of Z, and therefore there is (by Theorem 4.3) a Borel function U from Z to U such that $(\zeta,U(\zeta)) \in W$ for μ-a.a. $\zeta \in Z$. Thus $\int_Z^\oplus U(\zeta)d\mu(\zeta)$ is a unitary operator on $L^2(\mu;H)$ which is easily seen to intertwine $\int_Z^\oplus \pi(\zeta)d\mu(\zeta)$ and $\int_Z^\oplus \rho(\zeta)d\mu(\zeta)$.

Turning to the assertion concerning quasi-equivalence, one may evidently assume that $\pi(\zeta) \approx \rho(\zeta)$ for each $\zeta \in Z$. Put $H'(\zeta) = H(\zeta) \otimes \ell^2$ and $\pi'(\zeta)(R) = \pi(\zeta)(R) \otimes I$ for all $\zeta \in Z$ and $R \in R$. Then H' is a Borel field of Hilbert spaces on Z and by Lemma 7.4 there is a coherence α' for H' such that π' is an α'-Borel field of representations of R over H' and such that the two representations

$$\int_Z^\alpha \pi(\zeta)(\cdot)d\mu(\zeta) \otimes I$$

and

$$\int_Z^{\alpha'} \pi'(\zeta)d\mu(\zeta)$$

of R are equivalent. Thus $\int_Z^\alpha \pi(\zeta)d\mu(\zeta) \approx \int_Z^{\alpha'} \pi'(\zeta)d\mu(\zeta)$. If one defines K',ρ', and β' similarly then $\pi'(\zeta) \simeq \rho'(\zeta)$ for each $\zeta \in Z$ by Proposition A.3, and therefore $\int_Z^\alpha \pi(\zeta)d\mu(\zeta)$ and $\int_Z^\beta \rho(\zeta)d\mu(\zeta)$ are quasi-equivalent by the first part of the proof. □

Under the hypothesis that (in the notation of Theorem 12.1) $\pi(\zeta) \approx \rho(\zeta)$ for μ-a.a. $\zeta \in Z$ the preceeding theorem asserts the existence of a certain *-isomorphism of von Neumann algebras. In Section 28 it will be necessary to have more information about this *-isomorphism than is provided by Theorem 12.1; this additional information will be provided by Lemma 28.1.

COROLLARY 12.2. Let Z,μ,H, and α be as in Theorem 12.1, let π be an α-Borel field of representations of R [respectively, G] over H, and let π_0 be a representation of R [respectively, G] on a separable Hilbert space H_0. Then $\int_Z^\alpha \pi(\zeta)d\mu(\zeta)$ is equivalent to the representation $\pi_0(\cdot) \otimes I$ of R [respectively, G] on $H \otimes L^2(\mu)$ if $\pi_0 \simeq \pi(\zeta)$ for μ-a.a. $\zeta \in Z$ and quasi-equivalent to π_0 if $\pi_0 \approx \pi(\zeta)$ for μ-a.a. $\zeta \in Z$.

Proof. Put $K(\zeta) = H_0$ and $\rho(\zeta) = \pi_0$ for all $\zeta \in Z$, and let β be some constant coherence for K. Then ρ is certainly a β-Borel field of representations of R [respectively, G] over K and $\int_Z^\beta \rho(\zeta)d\mu(\zeta)$ is (by Proposition 5.2) equivalent to $\pi_0(\cdot) \otimes I$, where I is the identity operator on $L^2(\mu)$. This remark, the fact that $\pi_0 \approx \pi_0(\cdot) \otimes I$, and Theorem 12.1 clearly imply the corollary. □

THEOREM 12.3. Let Z be a Borel space, μ a Borel measure on Z, H a Borel field of Hilbert spaces on Z, α a coherence for H, and π_0 a representation of R [respectively, G] on $L^2(\mu;H,\alpha)$ such that $\pi_0(R)'$ [respectively, $\pi_0(G)'$] contains the algebra of diagonalizable operators on $L^2(\mu;H,\alpha)$. Then there is an α-Borel field of representations π of R [respectively, G] over H with $\pi_0 = \int_Z^\alpha \pi(\zeta)d\mu(\zeta)$. In particular, a given representation π_0 of R [respectively, G] on a separable Hilbert space always has a direct integral decomposition with respect to a given abelian von Neumann algebra contained in $\pi_0(R)'$ [respectively, $\pi_0(G)'$].

Proof. The second assertion follows from the first one and Theorem 9.1. In proving the first one it is enough, just as in the proof of Theorem 12.1, to consider the case of R. From the hypothesis and Theorem 7.1 one knows that $\pi_0(R)$ consists of decomposable operators, i.e., that for each R in R there is an α-Borel operator field $\hat{R}$ over H with $\pi_0(R) = \int_Z^\alpha \hat{R}(\zeta)d\mu(\zeta)$, and by Proposition B.1 and Theorem 7.1 these operator fields must be such that for given elements R and S in R and a given complex number a one has

$$(R^*)\hat{\,}(\zeta) = \hat{R}(\zeta)^*$$
$$(RS)\hat{\,}(\zeta) = \hat{R}(\zeta)\hat{S}(\zeta)$$
$$(aR)\hat{\,}(\zeta) = a\hat{R}(\zeta)$$
$$(R + S)\hat{\,}(\zeta) = \hat{R}(\zeta) + \hat{S}(\zeta)$$

for μ-a.a. $\zeta \in Z$.

Now let R_0 be a countable dense subset of R which is a *-algebra over $Q + iQ$. Then there must be a μ-null Borel set N in Z such that for each ζ in $Z - N$ the map $R \mapsto \hat{R}(\zeta)$ is a $(Q + iQ)$-linear norm-decreasing *-homomorphism of R_0 into $L(H(\zeta))$. So for each point ζ in $Z - N$ there is a representation $\pi(\zeta)$ of R on $H(\zeta)$ with $\pi(\zeta)(R) = \hat{R}(\zeta)$, $R \in R_0$; put $\pi(\zeta)(R) = 0$ for $\zeta \in Z$ $R \in R$. Then π is an α-Borel field of representations of R over H with $\pi_0 = \int_Z^\alpha \pi(\zeta)d\mu(\zeta)$. Indeed, take an R in R and a sequence (R_n) in R_0 converging to R. Then

$$\pi(\zeta)(R) = \lim_{n\to\infty} \pi(\zeta)(R_n) = \lim_{n\to\infty} \hat{R}_n(\zeta)$$

for each $\zeta \in Z - N$ (by Proposition B.1), hence $\pi(\cdot)(R)$ is an α-Borel operator field over H, and, moreover,

$$\int_Z^\alpha \pi(\zeta)(R)d\mu(\zeta) = \lim_{n\to\infty} \int_Z^\alpha \hat{R}_n(\zeta)d\mu(\zeta) = \pi_0(R)$$

by Theorem 7.1. □

THEOREM 12.4. Retain the notation and hypothesis of Theorem 9.2, and assume in addition that π and ρ are α-Borel and β-Borel fields of representations of either R on G over H and K, respectively, and that W intertwines $\int_Y^\beta \rho(\xi)d\nu(\xi)$ and $\int_X^\alpha \pi(\zeta)d\mu(\zeta)$. Then $V(\zeta)$ intertwines $\rho(\omega(\zeta))$ and $\pi(\zeta)$ for μ-a.a. $\zeta \in X$.

Proof. As in the proof of Theorem 12.1, it is enough to consider the case of R. For any R in R and any w in $L^2(\nu;K,\beta)$ one has

$$\int_X^\alpha \theta(\omega(\zeta))^{-\frac{1}{2}} \pi(\zeta)(R)V(\zeta)w(\omega(\zeta))d\mu(\zeta)$$

$$= \left[\int_X^\alpha \pi(\zeta)(R)d\mu(\zeta)\right] W \int_Y^\beta w(\xi)d\nu(\xi)$$

$$= W \int_Y^\beta \rho(\xi)(R)w(\xi)d\nu(\xi)$$

$$= \int_X^\alpha \theta(\omega(\zeta))^{-\frac{1}{2}} V(\zeta)\rho(\omega(\zeta))(R)w(\omega(\zeta))d\mu(\zeta)$$

and hence

$$\pi(\zeta)(R)V(\zeta)w(\omega(\zeta)) = V(\zeta)\rho(\omega(\zeta))(R)w(\omega(\zeta)) \tag{1}$$

for μ-a.a. $\zeta \in X$. Letting R_0 be as in the proof of Theorem 12.3 and W as in the proof of Theorem 9.2, there will be a μ-null Borel set P in X such that (1) holds for all $R \in R_0$, all $w \in W$, and all $\zeta \in Z - P$. But this implies that $V(\zeta)$ intertwines $\rho(\omega(\zeta))$ and $\pi(\zeta)$ for each $\zeta \in Z - P$. □

13. Maximal and Central Decompositions

Throughout this section R will denote a separable involutive Banach algebra, G a separable locally compact group, Z a standard Borel space, μ a Borel measure on Z, H a Borel field of Hilbert spaces on Z, α a coherence for H, M the algebra of diagonalizable operators on $L^2(\mu;H,\alpha)$, π an α-Borel field of representations of R [respectively, G] over H and ρ the representation $\int_Z^\alpha \pi(\zeta)d\mu(\zeta)$. Then $M \subset \rho(R)'$ [respectively, $M \subset \rho(G)'$], and this section is concerned with the relationship between the pair $M, \rho(R)'$ [respectively, $M, \rho(G)'$] and the properties of the $\pi(\zeta)$, $\zeta \in Z$. The proofs of the theorems of this section for the case of G can (as was explained in the proof of Theorem 12.1) be reduced to that of R, and hence only the latter will be given. Recall that a subset of Z is called μ-null if it is contained in a μ-null Borel subset of Z; actually, it will follow from Theorems 21.1(i) and 26.1 that the subsets of Z which occur in the statements of Theorems 13.1 and 13.2 are already Borel.

THEOREM 13.1. M is a maximal abelian subalgebra of $\rho(R)'$ [respectively, $\rho(G)'$] if and only if $\pi(\zeta)$ is irreducible for μ-a.a. $\zeta \in Z$.

Proof. Recall that M is maximal abelian in $\rho(R)'$ if and only if $M' \cap \rho(R)' \subset M$. Assume first that $\pi(\zeta)$ is irreducible for μ-a.a. $\zeta \in Z$, and consider an operator S in $M' \cap \rho(R)'$. Thus $S = \int_Z^\alpha A(\zeta) d\mu(\zeta)$ for some μ-essentially bounded α-Borel operator field A over H (by Theorem 7.1), and a by-now standard argument using a countable dense subset of R and Proposition B.1 will show that $A(\zeta) \in \pi(\zeta)(R)'$ for μ-a.a. $\zeta \in Z$, and hence that $S \in M$.

Now assume that, conversely, M is maximal abelian in $\rho(R)'$. In proving that $\pi(\zeta)$ is irreducible for μ-a.a. $\zeta \in Z$ one may (by formula (*) of Section 11) assume that H and α are constant. Thus each $\pi(\zeta)$ is a representation of R on a fixed separable Hilbert space H and $\pi(\cdot)(R)$ is a Borel function from Z to $L(H)$ for each $R \in R$. The set

$$W = \{(\zeta, A) \in Z \times L(H) : \|A\| \leq 1, A \notin C(H), \text{ and } A \in \pi(\zeta)(R)'\}$$

is a Borel (cf. the proof of Theorem 12.1), and hence a standard, subset of $Z \times L(H)$ and its image under the projection of $Z \times L(H)$ onto Z is the set

$$S = \{\zeta \in Z : \pi(\zeta) \text{ is not irreducible}\}$$

Thus S is an analytic subset of Z and so by Theorems 4.1 and 4.3 there are Borel sets R and T in Z and a Borel function A from R to $L(H)$ satisfying $R \subset S \subset T$, $\mu(T - R) = 0$, and $(\zeta, A(\zeta)) \in W$ for all $\zeta \in R$. Now if one puts $A(\zeta) = 0$ for $\zeta \in Z - R$ then $\int_Z^\alpha A(\zeta) d\mu(\zeta)$ lies in $M' \cap \rho(R)'$, and hence in M. But this means that $A(\zeta) \in C(H)$ for μ-a.a. $\zeta \in Z$, which can only occur if $\mu(R) = 0$, or equivalently, if $\mu(T) = 0$. □

THEOREM 13.2. (a) If $\pi(\zeta)$ is a factor representation for μ-a.a. $\zeta \in Z$ then M contains the center of $\rho(R)'$ [respectively, $\rho(G)'$].

13. MAXIMAL AND CENTRAL DECOMPOSITIONS

(b) If M is equal to the center of $\rho(R)'$, [respectively, $\rho(G)'$] then $\pi(\zeta)$ is a factor representation for μ-a.a. $\zeta \in Z$.

Proof. (a) Consider an operator S in $\rho(R)' \cap \rho(R)''$. Then S is decomposable as $\rho(R)'' \subset M'$ (Theorem 7.1), i.e., $S = \int_Z^\alpha A(\zeta) d\mu(\zeta)$ for some μ-essentially bounded α-Borel operator field A over H, and A must be such that $A(\zeta) \in \pi(\zeta)(R)'$ for μ-a.a. $\zeta \in Z$ (cf. the proof of Theorem 13.1). Next, one can use the Kaplansky density theorem (Theorem A.2) and Theorem 7.1 to find sequences (R_n) in R and (a_n) in C such that $\rho(R_n) + a_n I \to S$ strongly as $n \to \infty$ and $\pi(\zeta)(R_n) + a_n I \to A(\zeta)$ strongly as $n \to \infty$ for μ-a.a. $\zeta \in Z$. But then $A(\zeta)$ lies in the center of $\pi(\zeta)(R)''$ for μ-a.a. $\zeta \in Z$, and therefore S is diagonalizable, i.e., $S \in M$.

(b) Let S denote the norm-closed *-algebra of operators on $L^2(\mu;H,\alpha)$ generated by $\rho(R)$ and a sequence (A_n) which is weakly dense in the unit ball of $\rho(R)'$ (recall that $L^2(\mu;H,\alpha)$ is separable by Theorem 7.1 and that consequently such a sequence exists). Then S is a separable C*-algebra by Proposition B.1, and $S' = \rho(R)' \cap \rho(R)'' = M$. So one can apply Theorem 12.3 to the identity representation of S on $L^2(\mu;H,\alpha)$ to obtain an α-Borel field of representations σ of S over H with $A = \int_Z^\alpha \sigma(\zeta)(A) d\mu(\zeta)$ for each $A \in S$. Moreover, from Theorems 7.1 and 13.1, Proposition B.1, and the proof of Theorem 12.3 one knows that there is a μ-null Borel set N in Z such that $\sigma(\zeta)$ is irreducible, $\pi(\zeta) = \sigma(\zeta) \circ \rho$, and $\sigma(\zeta)(A_n) \in \pi(\zeta)(R)'$ for all n and all $\zeta \in Z - N$. But then

$$\pi(\zeta)(R)' \cap \pi(\zeta)(R)'' \subset \pi(\zeta)(R)' \cap \{\sigma(\zeta)(A_n): n \in \mathbb{N}\}'$$
$$= \sigma(\zeta)(S)'$$
$$= C(H(\zeta))$$

i.e., $\pi(\zeta)$ is a factor representation, whenever $\zeta \in Z - N$. □

In Example 15.3 it will be shown that the converse of part (a) of the preceeding theorem is false.

THEOREM 13.3. If M is contained in the center of $\rho(R)'$ [respectively, $\rho(G)'$] then there is a μ-null Borel set N in Z such that the $\pi(\zeta)$, $\zeta \in Z - N$, are mutually disjoint.

Proof. Take a sequence of Borel subsets (S_j) of Z separating the points of Z. Then for each j the operator $\int_Z^\alpha 1_{S_j}(\zeta) I_{H(\zeta)} d\mu(\zeta)$ lies in $\rho(R)''$, and so there are sequences $(R_{j,n})_{n\in\mathbb{N}}$ in R and $(a_{j,n})_{n\in\mathbb{N}}$ in $\mathbb{C}$ and a μ-null Borel set N_j in Z such that $\pi(\zeta)(R_{j,n}) + a_{j,n}I \to 1_{S_j}(\zeta)I$ strongly as $n \to \infty$ for each $\zeta \in Z - N_j$ (cf. the proof of Theorem 13.2). Now take two distinct points ζ and ζ' in $Z - \cup_j N_j$ and an operator T from $H(\zeta)$ to $H(\zeta')$ intertwining $\pi(\zeta)$ and $\pi(\zeta')$. Then there will be a j with, say, $\zeta \in S_j$ and $\zeta' \notin S_j$, and hence with

$$T = \lim_{n\to\infty} T[\pi(\zeta)(R_{j,n}) + a_{j,n}I]$$
$$= \lim_{n\to\infty} [\pi(\zeta')(R_{j,n}) + a_{j,n}I]T$$
$$= 0$$

Thus $N = \cup_j N_j$ will do. □

The following result (which will play a salient role in Section 28) is an obvious consequence of the last two theorems.

COROLLARY 13.4. If M is equal to the center of $\rho(R)'$ [respectively, $\rho(G)'$] then there is a μ-null Borel set N in Z such that the $\pi(\zeta)$, $\zeta \in Z - N$, are mutually disjoint factor representations.

14. Some Applications

PROPOSITION 14.1. The irreducible representations which act on separable Hilbert spaces of a separable locally compact group G separate the points of G.

Proof. If λ denotes the left regular representation of G and if r is some element of G other than the identity then (as is easily seen) $\lambda(r) \neq I$. So by applying Theorems 12.3 and 13.1 to λ and some maximal abelian subalgebra of $\lambda(G)'$ one will obtain an irreducible representation, say π, of G which acts on a separable Hilbert space and satisfies $\pi(r) \neq I$. □

PROPOSITION 14.2. If the representations of a separable involutive Banach algebra R separate the points of R then so do those irreducible representations of R which act on separable Hilbert spaces.

Proof. Let R be a nonzero element of R and let π be a representation of R on a Hilbert space H with $\pi(R) \neq 0$. Choose a vector x in H with $\pi(R)x \neq 0$,

15. EXAMPLES

and let E be the projection of H onto the closed subspace of H spanned by $\pi(R)x$. Then EH is nonzero and separable (by Proposition B.1), $E \in \pi(R)'$, and $\pi_E(R) \neq 0$. One can now apply to π_E the argument which was applied to λ in the proof of Proposition 14.1; this will yield an irreducible representation, say ρ, of R which acts on a separable Hilbert space and satisfies $\rho(R) \neq 0$. □

PROPOSITION 14.3. Suppose that R is a separable involutive Banach algebra and that π is a factor representation of R on a separable Hilbert space. Then there is an irreducible representation of R which acts on a separable Hilbert space and has the same kernel as π.

Proof. By Theorems 12.3 and 13.1 one may as well assume that $\pi = \int_Z^\alpha \rho(\zeta) d\mu(\zeta)$ for some standard Borel space Z, some Borel measure μ on Z, some Borel field of Hilbert spaces H on Z, some coherence α for H, and some α-Borel field of irreducible representations ρ of R over H. Consider a Borel subset S of Z with $\mu(S) > 0$, and put $E = \int_Z^\alpha 1_S(\zeta) \mathrm{Id}\mu(\zeta)$; then E is a nonzero projection in $\pi(R)'$, and hence $A \mapsto A_E$ is a *-isomorphism of $\pi(R)''$ onto $[\pi(R)'']_E$ by Proposition A.5. But then $\|A\| = \|A_E\|$ for all $A \in \pi(R)''$ by Proposition A.3, and hence

$$\|\pi(R)\| = \|\pi(R)E\| = \mu\text{-ess. sup } 1_S(\cdot)\|\rho(\cdot)(R)\|$$

for each $R \in R$ by Theorem 7.1. This implies that for each fixed element R in R the set

$$\{\zeta \in Z : |\ \|\rho(\zeta)(R)\| - \|\pi(R)\|\ | \geq \frac{1}{n}\}$$

is μ-null for $n = 1, 2, \ldots$, and consequently that $\|\rho(\zeta)(R)\| = \|\pi(R)\|$ for μ-a.a. $\zeta \in Z$. So if (R_n) is a dense sequence in R there will be a sequence (N_n) of μ-null Borel subsets of Z such that $\|\rho(\zeta)(R_n)\| = \|\pi(R)\|$ for all $\zeta \in Z - N_n$ and all n. But then $\|\rho(\zeta)(R)\| = \|\pi(R)\|$ for all $\zeta \in Z - \cup_n N_n$ and all $R \in R$, and therefore each of the representations $\rho(\zeta)$, $\zeta \in Z - \cup_n N_n$, has the same kernel as π. □

15. Examples

It seems easier to give examples of direct integral decompositions of representations of groups then of algebras, and that is what will be done in this section. In addition to illustrating the general theory just presented, the examples will exhibit some of the pathology and delineate some of the natural bounderies of direct integral theory. Naturally enough, the examples given here are based on those discussed in the previous chapter.

EXAMPLE 15.1. From Examples 10.4 to 10.6 one can easily compute the direct integral decomposition of the left regular representation λ of a separable locally compact abelian group G with respect to $\lambda(G)"$. Notice that this direct integral decomposition is a direct integral of irreducible representations, and hence that $\lambda(G)"$ is maximal abelian in $\lambda(G)'$ (Theorem 13.1). Actually, as the direct integral decomposition of $L^2(G)$ is a direct integral of one-dimensional Hilbert spaces one knows that $\lambda(G)"$ is a maximal abelian von Neumann algebra on $L^2(G)$, hence that $\lambda(G)" = \lambda(G)'$, and, finally, that there is essentially only one way of writing λ as a direct integral of irreducible representations. □

Notice that if G is the circle group then λ is actually a direct sum of irreducible representations and that if G is the real line then λ is a direct integral. The uniqueness result just described then implies that no direct sum of irreducible representations of $\mathbb{R}$ can be equivalent to the left regular representation of $\mathbb{R}$. In fact, the left regular representation of $\mathbb{R}$ is not equivalent to a representation of the form $\pi \oplus \rho$, where π is irreducible. Indeed, if this were the case then (as $\pi(\mathbb{R})"$ is abelian) π would have to act on a one-dimensional Hilbert space, and then (from Example 10.5) $L^\infty(\mathbb{R})$ would have to have a minimal projection. So direct integral theory really is needed if one wishes to build up arbitrary representations from irreducible ones.

EXAMPLE 15.2. This example will illustrate the remarkable degree of non-uniqueness associated with maximal direct integral decompositions. Roughly speaking, one can find two maximal direct integral decompositions of a certain representation such that the two associated families of irreducible representations have the following properties: every member of one family is disjoint from every member of the other family; one family consists of mutually inequivalent representations; and the other family can be chosen to either consist of mutually inequivalent representations or else to be such that each representation in it is equivalent to infinitely many other representations in it. The example is a continuation of Example 10.7 and

15. EXAMPLES

the notation introduced therein will be retained. It will, however, be convenient to put $\pi(\zeta)(a,b) = J_{(a,b)}(\zeta)$ and $\rho(\xi)(a,b) = K_{(a,b)}(\xi)$ for $\zeta \in X$, $\xi \in Y$, and $(a,b) \in G$. Then it follows easily from Example 10.7 that the $\pi(\zeta)$, $\zeta \in X$, and the $\rho(\xi)$, $\xi \in Y$, are unitary representations of G on $\ell^2(K)$ and $\ell^2(K^*)$, respectively, and that

(a) the direct integral decomposition of λ with respect to $\lambda'(K^* \times \{0\})''$ is $\int_X^\oplus \pi(\zeta) d\mu(\zeta)$, and

(b) the direct integral decomposition of λ with respect to $\lambda'(\{1\} \times K)''$ is $\int_Y^\oplus \rho(\xi) d\nu(\xi)$.

The two fields of unitary representations π and ρ will shortly be shown to have the following properties:

(1) the $\pi(\zeta)$, $\zeta \in X$, are irreducible and mutually disjoint,

(2) the $\rho(\xi)$, $\xi \in Y - \{1\}$, are irreducible and $\rho(1)$ is not irreducible (here 1 denotes the identity element in $\hat{K}$),

(3) for $\xi, \xi' \in Y$, $\rho(\xi)$ and $\rho(\xi')$ are equivalent if and only if $\xi = \xi'(a \cdot)$ for some a in K*, and

(4) $\pi(\zeta)$ and $\rho(\xi)$ are disjoint for all $\zeta \in X$ and all $\xi \in Y$.

Combining (a) and (b) with (1)-(4) and remembering Theorem 13.1 gives part of the nonuniqueness described above, and the rest of it comes about as follows.

As $\{srs^{-1}: s \in G\}$ is infinite whenever $r \neq (1,0)$ one knows that $\lambda'(G)'' = \lambda(G)'$ is a factor of type II_1 (it is, in fact, hyperfinite). Then as $\lambda'(K^* \times \{0\})''$ is maximal abelian in $\lambda(G)'$ a standard argument will show that, given a positive integer n, there are mutually orthogonal nonzero projections $E_1, \ldots, E_n$ in $\lambda'(K^* \times \{0\})''$ which are mutually equivalent in $\lambda(G)'$. Then there will be mutually disjoint Borel sets $S_1, \ldots, S_n$ in X such that VE_jV^* is the decomposable operator determined by 1_{S_j}, $1 \leq j \leq n$. Consider the subrepresentations $\lambda_j = \lambda_{E_j}$, $1 \leq j \leq n$, of λ : they are mutually equivalent; each one is quasi-equivalent to λ (use Proposition A.3); $[\lambda'(K^* \times \{0\})'']_{E_j}$ is contained in $\lambda_j(G)'$ and is *-isomorphic to $L^\infty(\mu_j)$, where μ_j is the restriction of μ to Borel subsets of S_j, $1 \leq j \leq n$; and, finally,

(c) the direct integral decomposition of λ_j with respect to $[\lambda'(K^* \times \{0\})'']_{E_j}$ is $\int_{S_j}^{\oplus} \pi(\zeta) d\mu_j(\zeta)$, $1 \leq j \leq n$.

The remaining part of the nonuniqueness assertion now comes from combining (1) and (c).

To prove (1), take two points ζ and ζ' in X and suppose that T is a bounded linear operator on $\ell^2(K)$ intertwining $\pi(\zeta)$ and $\pi(\zeta')$. Then

$$T\eta_b = \overline{\zeta(a)} T\pi(\zeta)(a,b)\eta_0$$
$$= \overline{\zeta(a)} \pi(\zeta')(a,b) T\eta_0$$
$$= \overline{\zeta(a)} \sum_{d \in K} <T\eta_0, \eta_d> \pi(\zeta')(a,b)\eta_d$$
$$= \sum_{d \in K} \overline{\zeta(a)} \zeta'(a) <T\eta_0, \eta_d> \eta_{ad+b}$$

and hence

$$|<T\eta_b, \eta_0>| = |<T\eta_0, \eta_{-ab}>|$$

for all $(a,b) \in G$. Now if $b \neq 0$ then $a \mapsto -ab$ is a one-one mapping of K^* onto itself, and thus $|<T\eta_0, \eta_1>| = |<T\eta_0, \eta_c>|$ for all $c \in K^*$. But this is possible only if $<T\eta_0, \eta_c> = 0$, $c \in K^*$, and so one must have $T\eta_0 = \alpha \eta_0$ for some complex number α. Now if $\zeta = \zeta'$ then $T\eta_b = \alpha \eta_b$ for all b in K by the above calculation. And if $\zeta \neq \zeta'$ then there must be an a in K^* with $\zeta(a) \neq \zeta'(a)$; but then

$$\alpha \eta_0 = T\eta_0 = \overline{\zeta(a)} \zeta'(a) \alpha \eta_0$$

(taking b = 0 in the first calculation), and hence $\alpha = 0$. This proves (1).

To prove (3), take two points ξ and ξ' in Y. If $\xi = \xi'(a \cdot)$ for some a in K^* then the image of a under the left regular representation of K^* is a unitary operator on $\ell^2(K^*)$ intertwining $\rho(\xi)$ and $\rho(\xi')$. Conversely, say that $\rho(\xi)$ and $\rho(\xi')$ fail to be disjoint, so that there is a nonzero bounded linear operator T on $\ell^2(K^*)$ intertwining $\rho(\xi)$ and $\rho(\xi')$. Now if $T\varepsilon_1 = 0$ then

$$T\varepsilon_a = T\rho(\xi)(a,0)\varepsilon_1 = \rho(\xi')(a,0)T\varepsilon_1 = 0$$

for all $a \in K^*$, contradicting the assumption that $T \neq 0$. Thus $<T\varepsilon_1, \varepsilon_a> \neq 0$ for some a in K^*, and then the calculation

15. EXAMPLES

$$\langle T\varepsilon_1, \varepsilon_a \rangle = \overline{\xi(b)} \langle T\rho(\xi)(1,b)\varepsilon_1, \varepsilon_a \rangle$$

$$= \overline{\xi(b)} \langle \rho(\xi')(1,b) T\varepsilon_1, \varepsilon_a \rangle$$

$$= \overline{\xi(b)} \langle T\varepsilon_1, \rho(\xi')(1,-b)\varepsilon_a \rangle$$

$$= \overline{\xi(b)} \xi'(a^{-1}b) \langle T\varepsilon_1, \varepsilon_a \rangle$$

which is valid for all b in K, shows that $\xi = \xi'(a^{-1}\cdot)$. This proves (3).

Turning to (2), it is obvious that $\rho(1)$ fails to be irreducible. Take a point ξ in $Y - \{1\}$ and consider an operator T in $\rho(\xi)(G)'$. If $\langle T\varepsilon_1, \varepsilon_a \rangle \neq 0$ for some a in $K^* - \{1\}$ then $\xi = \xi(a\cdot)$ from the calculation in the previous paragraph. But this is impossible, as it would imply that $\xi((1-a)b) = 1$ for all b in K, and hence that $\xi = 1$. Thus $T\varepsilon_1 = \alpha\varepsilon_1$ for some complex number α, and then

$$T\varepsilon_a = T\rho(\xi)(a,0)\varepsilon_1 = \rho(\xi)(a,0)T\varepsilon_1 = \alpha\varepsilon_a$$

for all a in K^*. This proves (2).

Finally, take a point ζ in X and a point ξ in Y, and suppose that T is a continuous linear operator from $\ell^2(K)$ into $\ell^2(K^*)$ intertwining $\pi(\zeta)$ and $\rho(\xi)$. One then has

$$|\langle T\eta_0, \varepsilon_a \rangle| = |\langle T\pi(\zeta)(c^{-1},0)\eta_0, \varepsilon_a \rangle|$$

$$= |\langle T\eta_0, \rho(\xi)(c,0)\varepsilon_a \rangle|$$

$$= |\langle T\eta_0, \varepsilon_{ac} \rangle|$$

for all a and c in K^*, which is impossible unless $T\eta_0 = 0$ (cf. the proof of (1)). But if $T\eta_0 = 0$ then one can easily show that $T = 0$. □

EXAMPLE 15.3. This example is a continuation of Example 10.8 and will provide the promised counter-example to the converse of Theorem 13.2 (a). Let H be an abelian subgroup of a countably infinite discrete group G, and let λ and λ' be the left and right regular representations of G, respectively, let $\mu, H, \alpha, (\delta_r)_{r \in G}, Z$, and W be as in Example 10.8 and put $[\pi(\zeta)(r)f](s) = f(r^{-1}s)$ for all $\zeta \in \hat{H}$, $r,s \in G$, and $f \in H(\zeta)$. Then π is an α-Borel field of representations of G over H, $W\lambda(\cdot)W^{-1} = \int_Z^\alpha \pi(\zeta)d\mu(\zeta)$, and $W\lambda'(H)''W^{-1}$ is the algebra of diagonalizable operators on $L^2(\mu; H, \alpha)$.

Suppose that there is an element t in G which is not in H but commutes with each element of H and satisfies $\lambda'(t) \in [\lambda(G) \cup \lambda'(H)]''$. Then $\lambda'(t)$ lies in $\lambda'(H)'$, and hence $W\lambda'(t)W^{-1}$ is decomposable; in fact, $W\lambda'(t)W^{-1} = \int_Z^\alpha A(\zeta)d\mu(\zeta)$, where, for each point ζ in Z, $A(\zeta)$ is the operator on $H(\zeta)$ defined by $[A(\zeta)f](r) = f(rt)$ for all $f \in H(\zeta)$ and $r \in G$. (Notice that $A(\zeta)$ is well-defined as t lies in the centralizer of H.) Then each $A(\zeta)$ lies in $\pi(\zeta)(G)'$ and fails to be a multiple of the identity. The later assertion comes from the fact that $(W\delta_e)(\zeta)(e) = 1$ while $[A(\zeta)(W\delta_e)(\zeta)](e) = (W\delta_e)(\zeta)(t) = 0$. There must be (using a by-now standard argument) a sequence (T_n) is the *-algebra generated by $\lambda(G) \cup \lambda'(H)$ converging strongly to $\lambda'(t)$. Now each $WT_n W^{-1}$ must be decomposable, and thus $WT_n W^{-1} = \int_Z^\alpha A_n(\zeta)d\mu(\zeta)$ for some μ-essentially bounded α-Borel operator field A_n over H, and one may evidently assume that $A_n(\zeta)$ lies in the *-algebra generated by $\pi(\zeta)(G)$, $\zeta \in Z$. Proposition 6.3 then implies that $A(\zeta) \in \pi(\zeta)(G)''$, and hence that $\pi(\zeta)$ fails to be a factor representation, for μ-a.a. $\zeta \in Z$.

Now let G be the free group on the two generators a and b and let H be the subgroup of G generated by a^2. Then $\lambda'(H)''$ properly contains the center of $\lambda(G)'$ (after all, $\lambda(G)'$ is a factor) and a commutes with each element of H. So to give a counter-example to the converse of Theorem 13.2 (a) it is enough (by the previous paragraph) to show that $\lambda'(a) \in [\lambda(G) \cup \lambda'(H)]''$. To do this, consider an operator B in $[\lambda(G) \cup \lambda'(H)]' = \lambda(G)' \cap \lambda'(H)'$. Then

$$\begin{aligned}\langle B\lambda'(a)\delta_r, \delta_s\rangle &= \langle B\lambda(ra^{-1})\delta_e, \delta_s\rangle \\ &= \langle \lambda(ra^{-1})B\delta_e, \delta_s\rangle \\ &= \langle B\delta_e, \delta_{ar^{-1}s}\rangle\end{aligned}$$

and

$$\begin{aligned}\langle \lambda'(a)B\delta_r, \delta_s\rangle &= \langle B\lambda(r)\delta_e, \delta_{sa}\rangle \\ &= \langle B\delta_e, \delta_{r^{-1}sa}\rangle\end{aligned}$$

for all r and s in G; to show, then, that B and $\lambda'(a)$ commute it is sufficient to show that $\langle B\delta_e, \delta_r\rangle = 0$ unless r is a power of a. But this follows from the calculation

$$\langle B\delta_e, \delta_r\rangle = \langle B\lambda(a^{2n})\lambda'(a^{2n})\delta_e, \delta_r\rangle$$

15. EXAMPLES

$$= \langle \lambda(a^{2n})\lambda'(a^{2n})B\delta_e, \delta_r \rangle$$

$$= \langle B\delta_e, \delta_{a^{-2n}ra^{2n}} \rangle$$

which is valid for all integers n, and the fact that the map $n \mapsto a^{-2n}ra^{2n}$ is one-one unless r is a power of a. □

EXAMPLE 15.4. This example, which will be the last one in this section, is a continuation of Example 10.9 and will answer the following natural question: how are the direct integral decompositions of a representation with respect to two abelian von Neumann algebras related if one of the two algebras in question is a subalgebra of the other? Retain the notation introduced in Example 10.9 and, in addition, let σ be a representation of a separable involutive Banach algebra R on H_0, and assume that $M \subset \sigma(R)'$. Then by Theorem 12.3 there is an α-Borel field of representations π of R over H with $U\sigma(\cdot)U^{-1} = \int_X^\alpha \pi(\zeta)d\mu(\zeta)$ as well as a β-Borel field of representations ρ of R over K with $V\sigma(\cdot)V^{-1} = \int_Y^\beta \rho(\xi)d\nu(\xi)$. For any element R in R and any two α-Borel vector fields v and w over H with the property that both $\|v(\cdot)\|$ and $\|w(\cdot)\|$ are bounded functions on Z the inner product

$$\langle \int_X^\alpha \pi(\zeta)(R)d\mu_\xi(\zeta) \int_X^\alpha v(\zeta)d\mu_\xi(\zeta), \int_X^\alpha w(\zeta)d\mu_\xi(v) \rangle$$

$$= \int_X \langle \pi(\zeta)(R)v(\zeta), w(\zeta) \rangle d\mu_\xi(\zeta)$$

is clearly a Borel function of ξ on Y. In view of Lemma 8.2 and the construction of the coherence γ, this means that $\xi \mapsto \int_X^\alpha \pi(\zeta)d\mu_\xi(\zeta)$ is a γ-Borel field of representations. But then one can form the representation $\int_Y^\gamma \int_X^\alpha \pi(\zeta)d\mu_\xi(\zeta)d\nu(\xi)$, and this representation is easily seen to be a direct integral decomposition σ with respect to N. In fact, the linear isometry WUV^{-1} of $\int_Y^\beta K(\xi)d\nu(\xi)$ onto $\int_Y^\gamma \int_X^\alpha H(\zeta)d\mu_\xi(\zeta)d\nu(\xi)$ intertwines $\int_Y^\beta \rho(\xi)d\nu(\xi)$ and $\int_Y^\gamma \int_X^\alpha \pi(\zeta)d\mu_\xi(\zeta)d\nu(\xi)$, and hence Theorem 12.4 implies that T(ξ) intertwines ρ(ξ) and $\int_X^\alpha \pi(\zeta)d\mu_\xi(\zeta)$ for ν-a.a. ξ ∈ Y. Thus one may as well identify the two fields ρ and $\xi \mapsto \int_X^\alpha \pi(\zeta)d\mu_\xi(\zeta)$.

The answer to the question posed in the previous paragraph is therefore formally the same as it was in Example 10.9, namely, that the larger the base space the finer the component representations will be. More precisely, in the above notation there is, for ν-a.a. ξ ∈ Y, an abelian von Neumann algebra acting on K(ξ) such that the direct integral decomposition of ρ(ξ) with respect to this algebra is $\int_X^\alpha \pi(\zeta)d\mu_\xi(\zeta)$; this abelian algebra will be identified in Example 20.2. □

HISTORICAL COMMENTS

The study of direct integrals of representations of locally compact groups and involutive Banach algebras was initiated by Godement and Mautner in the early 1950s [21, 43-45]. In particular, these authors proved Theorems 12.3 and 13.1. The part of Theorem 12.1 dealing with equivalence is due to Mackey [41, Theorem 10.1], and the part dealing with quasi-equivalence, to Ernest [18, Proposition 5]. Mautner proved Theorem 13.2 (b) [44, Lemma 1.2] and Mackey conjectured the special case of Theorem 13.3 in which ρ is multiplicity-free and M is the center of $\rho(R)'$ [41, p. 163]. This conjecture was verified by Guichardet [24], and Ernest [18, Proposition 3] showed that Guichardet's proof could actually be made to yield Theorem 13.3. At about the same time Naimark announced that he had obtained a proof of Theorem 13.3 [46, Theorem 1].

Proposition 14.3 is due to Dixmier and appears with two proofs, one of which is reproduced here, in [6, Corollaire 3 of Théorème 2]. Interestingly enough, for the other proof, which does not rely on direct integral theory, one must also assume separability of the algebra in question.

As was the case in the previous chapter, the examples contained in the present chapter are all well-known. Examples 15.2 and 15.4 are due to Mackey [38, Theorem 11; 39, Theorem 2.11], and Example 15.3 to Kadison [28, Section 3].

Chapter 4
DIRECT INTEGRALS OF VON NEUMANN ALGEBRAS

16. Hausdorff Metrics

Recall that a metric d on a set X is said to be *totally bounded* if it satisfied the following condition: given an $\varepsilon > 0$, there are finitely many points, say $x_1, \ldots, x_n$, in X with the property that $\min_{1 \leq i \leq n} d(x, x_i) < \varepsilon$ for each point x in X. The importance of such metrics resides in the fact that the completion of a metric space (X,d) is compact if and only if d is totally bounded.

Suppose that X is a metrizable topological space, and let $C(X)$ denote the set of all nonempty closed subsets of X. A metric d on X will be called *compatible* (with the given topology on X) if d induces the given topology on X and if the d-diameter of X is finite. Given a compatible metric d on X, put

$$d(x,F) = \inf\{d(x,y): y \in F\}$$

and

$$\hat{d}(F,G) = \max\{\sup_{x \in F} d(x,G), \sup_{y \in G} d(y,F)\}$$

for all $x \in X$ and all $F, G \in C(X)$. Thus $\hat{d}(F,G) \leq a$ if and only if it is the case that given $\varepsilon > 0$, there corresponds to each point x in F a point y in G with $d(x,y) \leq a + \varepsilon$, and vice-versa. Notice that $d(x,F) = 0$ if and only if $x \in F$, that $\hat{d}(F,G)$ is always finite and is zero if and only if $F = G$, and that $\hat{d}(F,G) = \hat{d}(G,F)$ for all $x \in X$ and $F, G \in C(X)$. If F, G, and H are elements in $C(X)$, if $x \in F$, and if $\varepsilon > 0$, then there is a point y in G with $d(x,y) \leq \hat{d}(F,G) + \varepsilon$ and, in turn, a point z in H with $d(y,z) \leq \hat{d}(G,H) + \varepsilon$; thus

$$d(x,H) \leq d(x,z) \leq \hat{d}(F,G) + \hat{d}(G,H) + 2\varepsilon$$

From this it follows readily that $\hat{d}$ satisfies the triangle inequality, and hence is a metric on $C(X)$; $\hat{d}$ is called the *Hausdorff metric on $C(X)$ associated with* d.

For example, let $X = (0,1]$ with the usual topology, let d be the usual metric on X, and put

$$e(x,y) = \min\{1, |x - y| + |\tfrac{1}{x} - \tfrac{1}{y}|\}, \quad x,y \in X$$

Then d and e are compatible metrics on X and $\hat{d}([\tfrac{1}{n},1],X) = \tfrac{1}{n}$ and $\hat{e}([\tfrac{1}{n},1],X) = 1$ for all n. This shows, in particular, that $\hat{d}$ and $\hat{e}$ induce distinct topologies on $C(X)$. Thus the topology on a metrizable space X does not in general determine a unique topology on $C(X)$. However, things are not as bad as they might be in that there is a uniquely determined compact metric topology on $C(X)$ if X is compact and a uniquely determined standard Borel structure on $C(X)$ if X is Polish. Before proving these two assertions it will be convenient to record an easily-verified property of Hausdorff metrics.

LEMMA 16.1. Let d be a compatible metric on a metrizable space X, let $F, F_1, F_2, \ldots$ be points in $C(X)$, and assume that $\hat{d}(F_n, F) \to 0$ as $n \to \infty$. If a convergent sequence (x_n) in X has the property that $x_n \in F_n$ for all n then its limit lies in F. Conversely, given a point in F, one can find a sequence (x_n) in X converging to it and satisfying $x_n \in F_n$ for each n.

PROPOSITION 16.2. If X is a compact metric space then the Hausdorff metrics associated with any two compatible metrics on X induce the same topology on $C(X)$, and this topology is compact.

Proof. Let d be any compatible metric on X. Given an $\varepsilon > 0$, there must be (as X is, by assumption, compact) a nonempty finite subset F_0 of X satisfying $d(x, F_0) < \varepsilon$ for all $x \in X$. But then for a given F in $C(X)$ one must have $\hat{d}(F,G) \leq \varepsilon$, where $G = \{x \in F_0 : d(x,F) < \varepsilon\}$. This shows that $\hat{d}$ is totally bounded.

Suppose that (F_n) is a $\hat{d}$-Cauchy sequence in $C(X)$, and consider the set $F = \cap_{n=1}^{\infty} \overline{(\cup_{k=n}^{\infty} F_k)}$. Then F is nonempty, for if one selects, for each n, a point x_n in F_n, then F will contain the (nonempty) set of limits of convergent subsequences of the sequence (x_n). Given an $\varepsilon > 0$, choose an integer

16. HAUSDORFF METRICS   65

N with the property that $\hat{d}(F_m,F_n) < \varepsilon$ whenever $m,n \geq N$, and let p be an integer satisfying $p \geq N$. If x is a point in F then there is an integer $k \geq N$ and a point y in F_k with $d(x,y) < \varepsilon$, and hence there is a point z in F_p with $d(x,z) < 2\varepsilon$. On the other hand, say that x is a point in F_p. Then for $k = 1,2,\ldots$ one can find a point x_k in F_{N+k} with $d(x,x_k) < \varepsilon$, and if y denotes the limit of some convergent subsequence of the sequence (x_k) then y must clearly lie in F and satisfy $d(x,y) < 2\varepsilon$. This proves that $\hat{d}(F,F_p) \leq 2\varepsilon$ whenever $p \geq N$, and consequently that $\lim_{n\to\infty} \hat{d}(F_n,F) = 0$.

Thus $\hat{d}$ is a complete and totally bounded metric, and therefore induces a compact topology, on $C(X)$. If e is a second compatible metric on X then $\hat{d}$ and $(d+e)\hat{}$ induce comparable, hence equal, compact metric topologies on $C(X)$. But then $\hat{d}$ and $\hat{e}$ must induce the same topology on $C(X)$ by symmetry. □

If X is a compact metric space the topology on $C(X)$ described in Proposition 16.2 will be called the *Hausdorff topology* on $C(X)$. The key to proving the promised result on Borel structures is the following consequence of the preceeding proposition.

LEMMA 16.3. If d is a totally bounded compatible metric on a Polish space X then the topology induced on $C(X)$ by $\hat{d}$ is Polish.

Proof. Let X* be the completion of (X,d) and let e be the natural metric on X*. Then X* is compact, and so (by Proposition 16.2) $\hat{e}$ induces a compact, hence Polish, topology on $C(X*)$. Regarding X as a subset of X* in the usual way, the map θ which associates to an F in $C(X)$ its closure in X* is clearly an isometry of $(C(X),\hat{d})$ into $(C(X*),\hat{e})$. Now as X is a Polish space there will be a sequence (V_n) of open sets in X* with $X = \cap_n V_n$ (Proposition 2.1), and then

$$\text{range}(\theta) = \cap_n \{F \in C(X*): F \cap V_n \text{ is dense in } F\}$$

Indeed, if an F in $C(X*)$ lies in range of θ then $F \cap X$ must be dense in F, and thus $F \cap V_n$ must be dense in F for each n. Conversely, if an F in $C(X*)$ is such that $F \cap V_n$ is dense in F for each n then (applying the Baire category theorem to X) $F \cap X$ will be dense in F and hence F will lie in the range of θ.

Give $C(X*)$ the topology induced by $\hat{e}$, give $C(X*)^2 = C(X*) \times C(X*)$ the product topology, and consider the sets

$$S_n = \{(F,G) \in C(X*)^2: F \cap V_n \subset G \subset F\}$$

Then each S_n is closed, for let $((F_k, G_k))$ be a sequence in S_n converging to a point (F,G) in $C(X^*)^2$. Given a point x in $F \cap V_n$, there will (by Lemma 16.1) be a sequence (x_k) converging to x and satisfying $x_k \in F_k \cap V_n$ for all sufficiently large k; but then $x_k \in G_k$ for all sufficiently large k, and hence $x \in G$. This shows that $F \cap V_n \subset G$, and a similar argument will show that $G \subset F$, and hence that $(F,G) \in S_n$. Now the diagonal D in $C(X^*)^2$ is also closed, and so (as $C(X^*)^2$ is compact and separable) each of the sets $S_n - D$ can be expressed as a countable union of compact sets. Letting π be the natural projection of $C(X^*)^2$ onto its first coordinate, each $\pi(S_n - D)$ will be an F_σ-subset of $C(X^*)$. Now it is easy to see that

$$\{F \in C(X^*): F \cap V_n \text{ is dense in } F\} = C(X^*) - \pi(S_n - D)$$

for each n. Thus range(θ) is a G_δ-subset of $C(X^*)$, and the desired conclusion now follows from Proposition 2.1. □

PROPOSITION 16.4. The Borel structures on $C(X)$ generated by the Hausdorff metrics associated with any two totally bounded compatible metrics on a Polish space X coincide, and this Borel structure is standard.

Proof. Suppose that d and e are two totally bounded compatible metrics on X. Then $d + e$ is certainly compatible, and it is even totally bounded. Indeed, if $d + e$ were not totally bounded there would be an $\varepsilon > 0$ and a sequence (x_n) in X such that $(d + e)(x_m, x_n) \geq \varepsilon$ whenever $m \neq n$. Now as d and e are totally bounded some subsequence of (x_n) must be d-Cauchy and must itself contain an e-Cauchy subsequence. But this is clearly a contradiction. Thus the Borel structures induced on $C(X)$ by $\hat{d}, \hat{e}$, and $(d + e)^\wedge$ are all standard by Lemma 16.3, and clearly those induced by $\hat{d}$ and $\hat{e}$ are smaller than that induced by $(d + e)^\wedge$. But then Corollary 2.10 implies that $\hat{d}$ and $\hat{e}$ induce the same Borel structures on $C(X)$. □

Given a Polish space X, one knows from the usual proof of the Urysohn metrization theorem that there is a totally bounded compatible metric d on X and from Proposition 16.4 that the Borel structure on $C(X)$ induced by $\hat{d}$ is standard and depends only on X. This Borel structure on $C(X)$ will be called the *Hausdorff Borel structure on* $C(X)$.

17. The Effros Borel Structure

This section will be concerned with the set vN(H) of all von Neumann algebras acting on some separable Hilbert space H. The object is to define a certain standard Borel structure, the so-called *Effros Borel structure*, on vN(H) and to derive a few simple properties of it. In doing so it will be convenient to consider a fixed separable Hilbert space H and to let L denote the set of all bounded linear operators on H, L_1 the unit ball of L equipped with the weak operator topology, and L_* the set of all ultraweakly continuous linear functionals on L. The most useful properties of the Effros Borel structure can be summarized as follows:

THEOREM 17.1. The Effros Borel structure on vN(H) is a standard Borel structure with the following properties:

(a) it is the weakest Borel structure on vN(H) making the functions $A \mapsto \|f|A\|$, $f \in L_*$, into the real line Borel, or equivalently, the weakest Borel structure making the map $A \mapsto A \cap L_1$ into $C(L_1)$ Borel, where $C(L_1)$ is given the Hausdorff Borel structure,

(b) there is a sequence (A_n) of Borel maps from vN(H) into L_1 with the property that for each A in vN(H) the sequence $(A_n(A))$ is weakly dense in the unit ball of A,

(c) there is a sequence (A_n) of Borel maps from vN(H) into L with the property that for each A in vN(H) the set $\{A_n(A): n \in \mathbb{N}\}$ is a *-algebra over $Q + iQ$ which is strongly dense in A,

(d) a map A from a Borel space Z into vN(H) is Borel if and only if there is a sequence (A_n) of Borel maps from Z into L with the property that for each point ζ in Z, $A(\zeta)$ is the von Neumann algebra generated by the operators $A_n(\zeta)$, $n \in \mathbb{N}$,

(e) $A \mapsto A'$ is a Borel isomorphism of vN(H), and

(f) $(A,B) \mapsto (A \cup B)''$ and $(A,B) \mapsto A \cap B$ are Borel maps from vN(H) $\times$ vN(H) into vN(H).

The proof of the theorem will not be taken up until a number of auxillary results have been established. As L_1 is a compact metric space one can give $C(L_1)$ the Hausdorff topology and Borel structure (see Section 16) and consider the set $C_c(L_1)$ consisting of the convex sets in $C(L_1)$ with the relative topology and Borel structure.

LEMMA 17.2. The map $S \mapsto S^* = \{S^*: S \in S\}$ is a homeomorphism of $C(L_1)$.

Proof. The map in question is clearly a bijection of order two of $C(L_1)$. To prove that it is continuous, take a sequence (x_n) which is strongly dense in the unit ball of H and put

$$d(A,B) = \sum_{m,n=1}^{\infty} 2^{-m-n} |<(A-B)x_m, x_n>|$$

for $A,B \in L_1$. Then d is a compatible metric on L_1 making the involution on L_1 isometric, and therefore the map in question is isometric with respect to the metric $\hat{d}$. □

LEMMA 17.3. There is a sequence (M_n) of continuous maps from $C_c(L_1)$ into L_1 with the property that for each set S in $C_c(L_1)$ the sequence $(M_n(S))$ is weakly dense in S.

Proof. Let (x_n) be a sequence of unit vectors in H which is strongly dense in the set of all unit vectors in H. Then given a nonzero operator A in L there will be at least one integer n with $<Ax_n, x_n> \neq 0$. (To see this simply recall that

$$4<Ax,y> = \sum_{k=0}^{3} i^k <A(x + i^k y), x + i^k y>$$

for all x,y in H.) So if one puts

$$<A,B> = \sum_{n=1}^{\infty} 2^{-n} <Ax_n, x_n><x_n, Bx_n>$$

and

$$d'(A,B) = \left[\sum_{n=1}^{\infty} 2^{-n} |<(A-B)x_n, x_n>|^2 \right]^{\frac{1}{2}}$$

for all $A,B \in L$ then $<\cdot,\cdot>$ is an inner product on L and d' is the metric associated with this inner product. Now d' is easily seen to induce the weak operator topology on bounded subsets of L, and hence the restriction d of $\frac{1}{2}d'$ to $L_1 \times L_1$ is a compatible metric on L_1. Now d is complete and hence there is, for each operator A in L_1 and each subset S of $C_c(L_1)$, a unique operator, say $M(A,S)$, in S satisfying $d(A,M(A,S)) = d(A,S)$. It is clear that if M were separately continuous on $L_1 \times C_c(L_1)$ then one could obtain the desired sequence of functions simply by taking a dense sequence (A_n) in L_1 and putting $M_n(S) = M(A_n, S)$ for all n and all S in $C_c(L_1)$.

To prove that M is separately continuous one first establishes an auxillary inequality. Consider a set S in $C_c(L_1)$ and operators A and S in

17. THE EFFROS BOREL STRUCTURE

L_1 and S, respectively, and put $T = M(A,S)$ and $a = d(A,S) - d(A,\mathcal{S})$. Then

$$\begin{aligned} d(S,T)^2 &= \tfrac{1}{4}\langle S - T, S - T\rangle \\ &= \tfrac{1}{2}\langle A - S, A - S\rangle + \tfrac{1}{2}\langle A - T, A - T\rangle - \langle A - \tfrac{1}{2}(S+T), A - \tfrac{1}{2}(S+T)\rangle \\ &\leq 2[d(A,S) + a]^2 + 2d(A,S)^2 - 4d(A,\mathcal{S})^2 \\ &= 4a\,d(A,S) + 2a^2 \\ &\leq 6a \end{aligned}$$

where the last inequality comes from the fact that the d-diameter of L_1 is 1, and thus

$$d(S,M(A,\mathcal{S}))^2 \leq 6[d(A,S) - d(A,\mathcal{S})] \qquad (1)$$

To prove the separate continuity of M consider sets $\mathcal{S}$ and $\mathcal{T}$ in $C_c(L_1)$ and operators A and B in L_1. Then

$$\begin{aligned} d(A,\mathcal{S}) &= d(M(A,\mathcal{S}),A) \\ &\geq d(M(A,\mathcal{S}),B) - d(B,A) \\ &\geq d(B,\mathcal{S}) - d(A,B) \end{aligned}$$

and consequently

$$d(B,\mathcal{S}) - d(A,\mathcal{S}) \leq d(A,B) \qquad (2)$$

Taking $S = M(B,\mathcal{S})$ in (1) and using (2) yields

$$\begin{aligned} d(M(A,\mathcal{S}), M(B,\mathcal{S}))^2 &\leq 6[d(A,M(B,\mathcal{S})) - d(A,\mathcal{S})] \\ &\leq 6[d(A,B) + d(B,\mathcal{S}) - d(A,\mathcal{S})] \\ &\leq 12\,d(A,B) \end{aligned}$$

On the other hand, given an $\varepsilon > 0$ there will be operators S and T in $\mathcal{S}$ and $\mathcal{T}$, respectively, satisfying

$$d(S,M(A,\mathcal{T})) \leq \hat{d}(\mathcal{S},\mathcal{T}) + \varepsilon$$

and

$$d(T,M(A,\mathcal{S})) \leq \hat{d}(\mathcal{S},\mathcal{T}) + \varepsilon$$

Then

$$d(A,S) \le d(A,M(A,T)) + d(M(A,T),S)$$
$$\le d(A,T) + \hat{d}(S,T) + \varepsilon$$
$$\le d(A,M(A,S)) + d(M(A,S),T) + \hat{d}(S,T) + \varepsilon$$
$$\le d(A,S) + 2\hat{d}(S,T) + 2\varepsilon$$

Now one can use this inequality and (1) to obtain

$$d(M(A,S), M(A,T)) \le d(M(A,S),S) + d(S,M(A,T))$$
$$\le 3[d(A,S) - d(A,S)]^{\frac{1}{2}} + \hat{d}(S,T) + \varepsilon$$
$$\le 3[2\hat{d}(S,T) + 2\varepsilon]^{\frac{1}{2}} + \hat{d}(S,T) + \varepsilon$$

and finally, as ε was arbitrary, that

$$d(M(A,S), M(A,T)) \le 6\hat{d}(S,T)^{\frac{1}{2}} + \hat{d}(S,T) \quad \square$$

COROLLARY 17.4. The given Borel structure on $C_c(L_1)$ is standard and is generated by the functions $S \mapsto \sup\{|f(S)|: S \in S\}$, $f \in L_*$, into the real line.

Proof. It follows easily from Lemma 16.1 that $C_c(L_1)$ is a closed subset of $C(L_1)$, and hence that the given Borel structure on $C_c(L_1)$ is standard. If (M_n) is as in Lemma 17.3 then

$$\sup\{|f(S)|: S \in S\} = \sup_n |f(M_n(S))|$$

for all f in L_* and all S in $C_c(L_1)$, and thus the functions in question are Borel. Now as the polars in L_* of distinct sets in $C_c(L_1)$ are distinct the functions $S \mapsto \sup\{|f(S)|: S \in S\}$, where f runs through some countable norm-dense subset of L_*, will separate the points of $C_c(L_1)$. The second assertion of the lemma therefore follows from Corollary 2.10. $\square$

COROLLARY 17.5. A map S from a Borel space Z into $C_c(L_1)$ is Borel if and only if there is a sequence (A_n) of Borel functions from Z to L_1 with the property that for each point ζ in Z the sequence $(A_n(\zeta))$ is weakly dense in $S(\zeta)$.

Proof. The necessity and sufficiency of the stated condition are obvious from Lemma 17.3 and Corollary 17.4, respectively. $\square$

17. THE EFFROS BOREL STRUCTURE

It will now be necessary to introduce some more notation. Let $W(L)$ denote the set of all ultraweakly closed subspaces of L equipped with the weakest Borel structure making the map $S \mapsto S \cap L_1$ into $C_c(L_1)$ Borel. Recall that L_* is actually a separable Banach space, and hence that one can consider the Hausdorff Borel structure (see Section 16) on $C(L_*)$; let $N(L_*)$ be the Borel subspace of $C(L_*)$ based on the set of norm-closed subspaces of L_*. Recall also that the natural bilinear form on $L_* \times L$ induces an isometric isomorphism between L and the dual of L_* and that the corresponding weak *-topology on L is just the ultraweak topology. This means, in particular, that the map $N \mapsto N^o$ which sends an N in $N(L_*)$ into its polar in L is one-one and onto $W(L)$.

LEMMA 17.6. *The given Borel structure on $N(L_*)$ is standard and is generated by the functions $N \mapsto \|f + N\|$, $f \in L_*$, into the real line (here $\|f + N\|$ denotes, as usual, the norm of the image of f in the quotient space L_*/N).*

Proof. Let d be any totally bounded compatible metric on L_* and give $C(L_*)$ the topology induced by $\hat{d}$. Then it follows easily from Lemma 16.1 that $N(L_*)$ is a closed subset of $C(L_*)$, and hence that $N(L_*)$ is a standard Borel space.

Say that a functional f in L_* and a nonnegative number a are given. If a sequence (N_k) in $N(L_*)$ converges to an N in $N(L_*)$ and satisfies $\|f + N_k\| \geq a$ for each k then $\|f + N\| \geq a$. Indeed, given a functional g in N, there will be functionals g_k in N_k satisfying $d(g, g_k) \leq \hat{d}(N, N_k) + 2^{-k}$, hence satisfying $\|g - g_k\| \to 0$ as $k \to \infty$, and thus

$$\|f + g\| \geq \lim\sup\nolimits_{k \to \infty} \|f + g_k\| - \|g - g_k\| \geq a$$

This proves that the functions in question are Borel. Now the functions $N \mapsto \|f + N\|$, where f runs through some countable norm-dense subset of L_*, separate the points of L_* and the second assertion of the lemma therefore follows from Corollary 2.10. □

COROLLARY 17.7. *The map $N \mapsto N^o$ is a Borel isomorphism of $N(L_*)$ onto $W(L)$.*

Proof. That the map in question is bijective has already been pointed out. That it is a Borel isomorphism follows from Corollary 17.4, Lemma 17.6, and the fact that $\|f + N\| = \|f|N^o\|$ for $f \in L_*$ and $N \in N(L_*)$. □

LEMMA 17.8. *The map $S \mapsto S'$ from $W(L)$ into itself is Borel.*

Proof. Applying Corollary 17.5 to the identity map on $W(L)$ gives a sequence (A_n) of Borel maps from $W(L)$ to L_1 such that for each S in $W(L)$ the sequence $(A_n(S))$ is weakly dense in the unit ball of S. Thus

$$S' = \{B \in L : A_n(S)B - BA_n(S) = 0 \text{ for } n \in \mathbb{N}\} \tag{3}$$

for $S \in W(L)$.

Let M be the set of all those sequences (B_n) in L satisfying

$$\|(B_n)\| = \sup_n \|B_n\| < \infty \tag{4}$$

and M_* the set of all those sequences (f_n) in L_* satisfying

$$\|(f_n)\| = \sum_{n=1}^{\infty} \|f_n\| < \infty \tag{5}$$

Then M and M_* are Banach spaces with respect to the coordinatewise linear operations and the norms defined by (4) and (5), respectively, and the bilinear form $\langle \cdot, \cdot \rangle$ on $M_* \times M$ given by

$$\langle (f_n), (B_n) \rangle = \sum_{n=1}^{\infty} f_n(B_n)$$

induces an isometric isomorphism between M and the dual of M_*. For each S in $W(L)$ one can define a norm-continuous linear map $S(S)$ from M_* into L^* (L^* is the dual of the Banach space L) by putting

$$S(S)((f_n))(B) = \sum_{n=1}^{\infty} f_n(A_n(S)B - BA_n(S))$$

The range of each $S(S)$ is actually contained in L_*. To see this, fix a subspace S in $W(L)$ and a sequence (f_n) in M_*, and define a sequence (g_n) in L_* by putting

$$g_n(B) = f_n(A_n(S)B - BA_n(S)), \quad B \in L$$

Then

$$\left\| S(S)((f_n)) - \sum_{n=1}^{N} g_n \right\| = \sup\left\{ \left\| \sum_{n=N+1}^{\infty} g_n(B) \right\| : B \in L_1 \right\}$$

$$\leq 2 \sum_{n=N+1}^{\infty} \|f_n\|$$

and therefore $S(S)((f_n))$ is a norm-limit of functionals in L_* and so must

17. THE EFFROS BOREL STRUCTURE

itself lie in L_*. The dual $S(S)^*$ of $S(S)$, where $S(S)$ is regarded as a mapping from M_* to L_*, is the mapping from L into M given by

$$S(S)^*(B)_n = A_n(S)B - BA_n(S), \quad B \in L$$

From (3) one knows that

$$S' = \text{kernel } S(S)^*, \quad S \in W(L) \tag{6}$$

and from the general theory of linear operators that

$$(\text{kernel } S(S)^*)^o = (\text{range } S(S))^- \tag{7}$$

Now take a weakly dense sequence (B_j) in L_1 and a norm-dense sequence (h_k) in M_*, and write $h_k = (f_{k,n})_{n \in \mathbb{N}}$ for each k. Then using (6) and (7), one has that for any f in L_* and any S in $W(L)$,

$$\|f|S'\| = \|f + (\text{kernel } S(S)^*)^o\|$$
$$= \inf_k \|f + S(S)(h_k)\|$$
$$= \inf_k \sup_j |f(B_j) + \sum_{n=1}^{\infty} f_{k,n}(A_n(S)B_j - B_j A_n(S))|$$

An application of Corollary 17.4 will now complete the proof of the lemma. $\square$

One can now give the proof of Theorem 17.1. It is obvious that Corollary 17.4 implies that the two Borel structures described in (a) actually coincide, that Corollary 17.5 implies (b) and the "only if" part of (d), and that Lemma 17.8 implies (e). It is also obvious that (b) and (d) together imply that the first of the two maps mentioned in (f) is Borel, and that (e) then implies that the second map too is Borel. To prove (c), let (A_n) be the sequence given by (b) and let $B_1, B_2, \ldots$ be an enumeration of the elements of the *-algebra over $Q + iQ$ of functions from $vN(H)$ to L generated by $A_1, A_2, \ldots$ Then for each A in $vN(H)$ the strong closure of $\{B_n(A): n \in \mathbb{N}\}$ is convex, hence weakly closed, and is therefore all of A. The Borel structure on $vN(H)$ defined in (a) is standard since $vN(H)$ is a Borel subset of $W(L)$ by Lemmas 17.2 and 17.8 and since $W(L)$ is itself a standard Borel space by Lemma 17.6 and Corollary 17.7. Finally, to prove the "if" part of (d), suppose that A is a map from a Borel space Z into $vN(H)$ and that there is a sequence (A_n) of Borel maps from Z into L such that for each point ζ in Z, the operators $A_n(\zeta)$, $n \in \mathbb{N}$, generate $A(\zeta)$ as a von Neumann algebra. With no loss of generality, one may assume that the set $\{A_n(\zeta): n \in \mathbb{N}\}$ is a *-algebra over $Q + iQ$ for each point ζ in Z. Putting $B_n(\zeta) = A_n(\zeta)$ if $\|A_n(\zeta)\| \leq 1$ and $B_n(\zeta) = 0$ otherwise, one obtains a sequence (B_n) of Borel maps from Z into L_1. For each point ζ in Z the norm-closure $B(\zeta)$ of $\{A_n(\zeta): n \in \mathbb{N}\}$ is a *-algebra of operators on H which is weakly dense in $A(\zeta)$, and the sequence $(B_n(\zeta))$ is norm-dense in the unit ball of $B(\zeta)$ and hence (by the Kaplansky density theorem (Theorem A.2)) weakly dense in the unit ball of $A(\zeta)$. But this is, by Corollary 17.5, enough to show that A is Borel.

18. Definitions and Some Elementary Properties

Consider a Borel space Z, a Borel field of Hilbert spaces H on Z, and a coherence α for H. A *field of von Neumann algebras over* H is by definition a function associating with each point ζ in Z a von Neumann algebra acting on $H(\zeta)$. A field of von Neumann algebras A over H will be called α-*Borel* if the map $\zeta \mapsto \alpha(\zeta) A(\zeta) [\alpha(\zeta)* | \ell_n^2]$ from $Z_n = \{\zeta \in Z: \dim H(\zeta) = n\}$ with the relative Borel structure to $vN(\ell_n^2)$ is Borel for each n, $0 \leq n \leq \infty$, or equivalently (see Theorem 17.1), if there is a sequence (A_n) of α-Borel operator fields over H with the property that the sequence $(A_n(\zeta))$ generates $A(\zeta)$ for each $\zeta \in Z$; in the latter case the sequence (A_n) will be called an α-*Borel generating sequence for* A. Quite clearly, if A is an α-Borel field of von Neumann algebras over H then one can find an α-Borel generating sequence (A_n) for A satisfying any one of the following three conditions: $-I \leq A_n(\zeta) \leq I$ for all n and all ζ; for each ζ the set $\{A_n(\zeta): n \in \mathbb{N}\}$ is a *-algebra over $Q + iQ$ which is strongly dense in $A(\zeta)$; or, for each ζ the set $\{A_n(\zeta): n \in \mathbb{N}\}$ is a group of unitary operators in $A(\zeta)$. Notice that (by Theorem 17.1) if $A, A_1, A_2, \ldots$ are α-Borel fields of von Neumann algebras over H then so are $A(\cdot)'$, $A(\cdot) \cap A(\cdot)'$, $[\cup_{n=1}^{\infty} A_n(\cdot)]''$, and $\cap_{n=1}^{\infty} A_n(\cdot)$.

For example, $\zeta \mapsto C(H(\zeta))$ and $\zeta \mapsto L(H(\zeta))$ are α-Borel fields of von Neumann algebras over H. A less trivial example is $\zeta \mapsto \pi(\zeta)(R)''$ [respectively, $\zeta \mapsto \pi(\zeta)(G)''$], where π is an α-Borel field of representations of a separable involutive Banach algebra R [respectively, of a separable locally compact group G] over H. The following lemma shows that every α-Borel field of von Neumann algebras over H actually has this form.

LEMMA 18.1. Let Z, H, and α be as above, let A be an α-Borel field of von Neumann algebras over H, and let G be the free group on infinitely many gen-

18. DEFINITIONS AND SOME ELEMENTARY PROPERTIES

erators. Then there is an α-Borel field of representations π of G over H such that $\pi(\zeta)(G)'' = A(\zeta)$ for each ζ in Z.

Proof. Choose an α-Borel generating sequence (A_n) for A such that $\{A_n(\zeta): n \in \mathbb{N}\}$ is a group of unitary operators for each $\zeta \in Z$. Letting $r_1, r_2, \ldots$ be the free generators for G, there is for each point ζ in Z a unique representation $\pi(\zeta)$ of G on $H(\zeta)$ with $\pi(\zeta)(r_n) = A_n(\zeta)$ for each $n \in \mathbb{N}$. But then π is obviously α-Borel and satisfies $\pi(\zeta)(G)'' = A(\zeta)$, $\zeta \in Z$. □

Let Z, H, and α be as above, let μ be a Borel measure on Z, and let A be an α-Borel field of von Neumann algebras over H. The set consisting of all those operators on $L^2(\mu; H, \alpha)$ of the form $\int_Z^\alpha A(\zeta) d\mu(\zeta)$, where A is some μ-essentially bounded α-Borel operator field over H with $A(\zeta) \in A(\zeta)$ for μ-a.a. $\zeta \in Z$, will be denoted by $\int_Z^\alpha A(\zeta) d\mu(\zeta)$ or, in case H and ζ are constant, by $\int_Z^\oplus A(\zeta) d\mu(\zeta)$. Then $\int_Z^\alpha A(\zeta) d\mu(\zeta)$ is certainly a *-algebra of operators on $L^2(\mu; H, \alpha)$ whose center contains the algebra of diagonalizable operators. For example, $\int_Z^\alpha C(H(\zeta)) d\mu(\zeta)$ and $\int_Z^\alpha L(H(\zeta)) d\mu(\zeta)$ are just the algebras of diagonalizable and decomposable operators on $L^2(\mu; H, \alpha)$, respectively, and if Z is discrete and if μ is counting measure on Z then $\int_Z^\alpha A(\zeta) d\mu(\zeta)$ is nothing but $\oplus_{\zeta \in Z} A(\zeta)$.

THEOREM 18.2. *Let Z, μ, H, α and A be as above, and assume that μ is standard. Then $\int_Z^\alpha A(\zeta) d\mu(\zeta)$ is a von Neumann algebra and is, in fact, generated by the diagonalizable operators on $L^2(\mu; H, \alpha)$ together with the $\int_Z^\alpha A_n(\zeta) d\mu(\zeta)$, where (A_n) is any α-Borel generating sequence for A.*

Proof. Let M denote the algebra of diagonalizable operators on $L^2(\mu; H, \alpha)$ and B_0 and B the *-algebra and the von Neumann algebra, respectively, generated by M and the $\int_Z^\alpha A_n(\zeta) d\mu(\zeta)$, $n \in \mathbb{N}$. Let S be an element of B'. Then S must be decomposable, i.e., $S = \int_Z^\alpha B(\zeta) d\mu(\zeta)$ for some μ-essentially bounded α-Borel operator field B over H, and $B(\zeta)$ must commute with each $A_n(\zeta)$ and with each $A_n(\zeta)^*$ for μ-a.a. $\zeta \in Z$. Thus $B(\zeta) \in A(\zeta)'$ for μ-a.a. $\zeta \in Z$, and hence $S \in [\int_Z^\alpha A(\zeta) d\mu(\zeta)]'$. This shows that $\int_Z^\alpha A(\zeta) d\mu(\zeta) \subset B$.

In proving the reverse inclusion, first notice that $M \subset B'$, and hence that B consists of decomposable operators. So if T is an operator in B then (by Theorem 7.1 and the Kaplansky density theorem (Theorem A.2)) there will be μ-essentially bounded α-Borel operator fields $B, B_1, B_2, \ldots$ over H with $T = \int_Z^\alpha B(\zeta) d\mu(\zeta)$, with $\int_Z^\alpha B_n(\zeta) d\mu(\zeta) \in B_0$ for all n, and with $B_n(\zeta) \to B(\zeta)$ strongly as $n \to \infty$ for μ-a.a. $\zeta \in Z$. But then $B(\zeta) \in A(\zeta)$ for μ-a.a. $\zeta \in Z$ as $B_0 \subset \int_Z^\alpha A(\zeta) d\mu(\zeta)$, and therefore $T \in \int_Z^\alpha A(\zeta) d\mu(\zeta)$. □

With Z, μ, H, α and A as in the above theorem, the von Neumann algebra $\int_Z^\alpha A(\zeta)d\mu(\zeta)$ is called the *direct integral of A with respect to μ and α*. This algebra will also be called a *direct integral decomposition* of a von Neumann algebra B with respect to a subalgebra M of the center of B if there is a linear isometry W from the Hilbert space on which B acts onto $L^2(\mu;H,\alpha)$ such that $WBW^{-1} = \int_Z^\alpha A(\zeta)d\mu(\zeta)$ and $WMW^{-1} = \int_Z^\alpha C(H(\zeta))d\mu(\zeta)$; and, moreover, the decomposition will be called *central* if M is the center of B. The next result (which follows immediately from Theorem 18.2) says that $\int_Z^\alpha A(\zeta)d\mu(\zeta)$ essentially determines A.

COROLLARY 18.3. Let Z, μ, H, and α be as in Theorem 18.2, and let A and B be two α-Borel fields of von Neumann algebras over H. Then $\int_Z^\alpha A(\zeta)d\mu(\zeta) \subset \int_Z^\alpha B(\zeta)d\mu(\zeta)$ if and only if $A(\zeta) \subset B(\zeta)$ for μ-a.a. $\zeta \in Z$.

Let Z, μ, H, α, and A continue to be as in Theorem 18.2, let $Z_n = \{\zeta \in Z: \dim H(\zeta) = n\}$, $n = \infty, 0, 1, 2, \ldots$, let μ_n be the restriction of μ to the Borel subsets of Z_n, and let W be the linear isometry of $L^2(\mu;H,\alpha)$ onto $\oplus_{0 \leq n \leq \infty} L^2(\mu_n; \ell_n^2)$ constructed in Section 7. Then

$$W\left[\int_Z^\alpha A(\zeta)d\mu(\zeta)\right]W^{-1} = \oplus_{0 \leq n \leq \infty} \int_{Z_n}^\oplus \alpha(\zeta)A(\zeta)[\alpha(\zeta)*|\ell_n^2]d\mu_n(\zeta) \qquad (*)$$

This formula will, of course, play a role similar to that of formula (*) of Section 11.

19. Some Further Properties

Throughout this section Z will denote a standard Borel space, μ a Borel measure on Z, H a Borel field of Hilbert spaces on Z, α a coherence for H, and M the algebra of diagonalizable operators on $L^2(\mu;H,\alpha)$.

THEOREM 19.1. If A is an α-Borel field of von Neumann algebras over H then so are $A(\cdot)'$ and $A(\cdot) \cap A(\cdot)'$, and

$$\int_Z^\alpha A(\zeta)'d\mu(\zeta) = \left[\int_Z^\alpha A(\zeta)d\mu(\zeta)\right]'$$

and

$$\int_Z^\alpha [A(\zeta) \cap A(\zeta)']d\mu(\zeta) = \left[\int_Z^\alpha A(\zeta)d\mu(\zeta)\right] \cap \left[\int_Z^\alpha A(\zeta)d\mu(\zeta)\right]'$$

19. SOME FURTHER PROPERTIES

Proof. That the two fields in question are α-Borel was pointed out in Section 18. An operator S in $\left[\int_Z^\alpha A(\zeta)d\mu(\zeta)\right]'$ must be decomposable, i.e., must be of the form $S = \int_Z^\alpha B(\zeta)d\mu(\zeta)$ for some μ-essentially bounded α-Borel operator field B over H. If (A_n) is an α-Borel generating sequence for A with $-I \leq A_n(\zeta) \leq I$ for all n and ζ then S must commute with each $\int_Z^\alpha A_n(\zeta)d\mu(\zeta)$, and thus $B(\zeta)$ must commute with each $A_n(\zeta)$ for μ-a.a. $\zeta \in Z$. But this means that $B(\zeta) \in A(\zeta)'$ for μ-a.a. $\zeta \in Z$, i.e., that $S \in \int_Z^\alpha A(\zeta)'d\mu(\zeta)$. This proves that $\left[\int_Z^\alpha A(\zeta)d\mu(\zeta)\right]' \subset \int_Z^\alpha A(\zeta)'d\mu(\zeta)$; the other inclusion is obvious. Finally, the second formula is an easy consequence of the first. □

COROLLARY 19.2. *If $A_1, A_2, \ldots$ are α-Borel fields of von Neumann algebras over H then so are $[\cup_{n=1}^\infty A_n(\cdot)]''$ and $\cap_{n=1}^\infty A_n(\cdot)$, and*

$$\int_Z^\alpha \left[\cup_{n=1}^\infty A_n(\zeta)\right]'' d\mu(\zeta) = \left[\cup_{n=1}^\infty \int_Z^\alpha A_n(\zeta)d\mu(\zeta)\right]''$$

and

$$\int_Z^\alpha \left[\cap_{n=1}^\infty A_n(\zeta)\right] d\mu(\zeta) = \cap_{n=1}^\infty \int_Z^\alpha A_n(\zeta)d\mu(\zeta)$$

Proof. Again, that the two fields in question are α-Borel was pointed out in Section 18. The second of the two formulae in question is obviously true, and the first one follows easily from the second one and the theorem. □

Consider a separable involutive Banach algebra R, an α-Borel field of representations π of R over H, and the associated representation $\rho = \int_Z^\alpha \pi(\zeta)d\mu(\zeta)$ of R on $L^2(\mu;H,\alpha)$ and α-Borel field of von Neumann algebras $\pi(\cdot)(R)''$ over H. Then it is certainly true that $\rho(R)'' \subset \int_Z^\alpha \pi(\zeta)(R)''d\mu(\zeta)$ and that $M \subset \rho(R)'$; and as $\int_Z^\alpha \pi(\zeta)(R)''d\mu(\zeta) = [M \cup \rho(R)]''$ by Theorem 18.2, it is easy to see that

$$\int_Z^\alpha \pi(\zeta)(R)''d\mu(\zeta) \subset \rho(R)''$$

if and only if $M \subset \rho(R)$. These remarks can be summarized as follows:

PROPOSITION 19.3. *If R and π are as above then the following are equivalent:*

(1) $\left[\int_Z^\alpha \pi(\zeta)d\mu(\zeta)\right](R)'' = \int_Z^\alpha \pi(\zeta)(R)''d\mu(\zeta)$

(2) M *is contained in the center of* $\left[\int_Z^\alpha \pi(\zeta)d\mu(\zeta)\right](R)''$

The remaining results from this section are all analogues of ones in Chapters 2 and 3. They are useful but not surprising, and no new techniques appear in their proofs.

THEOREM 19.4. If B is a von Neumann algebra on $L^2(\mu;H,\alpha)$ whose center contains M then there is an α-Borel field of von Neumann algebras A over H with $B = \int_Z^\alpha A(\zeta)d\mu(\zeta)$. In particular, a given von Neumann algebra always has a direct integral decomposition with respect to a given subalgebra of its center.

Proof. The second statement follows from the first one together with Theorem 9.1. To prove the first, let G be the free group on infinitely many generators and let ρ be a unitary representation of G on $L^2(\mu;H,\alpha)$ with $\rho(G)'' = B$. Then by Theorem 12.3 there is an α-Borel field of representations π of G over H such that $\rho = \int_Z^\alpha \pi(\zeta)d\mu(\zeta)$. The theorem now follows from Proposition 19.3. □

THEOREM 19.5. Let $X = Z$, let Y, ν, K, β, N, W, V, and ω be as in Theorem 9.2, and assume that A and B are α-Borel and β-Borel fields of von Neumann algebras over H and K, respectively, and that

$$W\left[\int_Y^\beta B(\xi)d\nu(\xi)\right]W^{-1} = \int_X^\alpha A(\zeta)d\mu(\zeta)$$

Then $V(\zeta)B(\omega(\zeta))V(\zeta)^{-1} = A(\zeta)$ for μ-a.a. $\zeta \in X$.

Proof. Let G be the free group on infinitely many generators and let π be an α-Borel field of representations of G over H such that $\pi(\zeta)(G)'' = A(\zeta)$, $\zeta \in X$, (cf. the proof of Lemma 18.1) and put $\hat{\pi} = \int_X^\alpha \pi(\zeta)d\mu(\zeta)$ and $\hat{\rho} = W^{-1}\hat{\pi}(\cdot)W$. Then $\hat{\rho}$ is a representation of G on $L^2(\nu;K,\beta)$ with $N \subset \hat{\rho}(G)'$, and thus $\hat{\rho} = \int_Y^\beta \rho(\xi)d\nu(\xi)$ for some β-Borel field of representations ρ of G over K (Theorem 12.3). But then $\rho(\cdot)(G)''$ is a β-Borel field of von Neumann algebras over K satisfying

$$\int_Y^\beta \rho(\xi)(G)''d\nu(\xi) = \left[N \cup \hat{\rho}(G)\right]''$$
$$= W^{-1}\left[M \cup \hat{\pi}(G)\right]''W$$
$$= W^{-1}\left[\int_X^\alpha A(\zeta)d\mu(\zeta)\right]W$$
$$= \int_Y^\beta B(\xi)d\nu(\xi)$$

by Theorem 18.2, and therefore $B(\cdot) = \rho(\cdot)(G)''$ ν-a.e. by Corollary 18.3. An application of Theorem 12.4 will now complete the proof. □

19. SOME FURTHER PROPERTIES

THEOREM 19.6. If A is an α-Borel field of von Neumann algebras over H then M is equal to the center of $\int_Z^\alpha A(\zeta) d\mu(\zeta)$ if and only if $A(\zeta)$ is a factor for μ-a.a. $\zeta \in Z$.

Proof. Let ρ be a representation of the free group on infinitely many generators G on $L^2(\mu;H,\alpha)$ such that $\rho(G)'' = \int_Z^\alpha A(\zeta) d\mu(\zeta)$. Then (by Theorem 12.3) there is an α-Borel field of representations π of G over H such that $\rho = \int_Z^\alpha \pi(\zeta) d\mu(\zeta)$, and, moreover, $\pi(\zeta)(G)'' = A(\zeta)$ for μ-a.a. $\zeta \in Z$ by Proposition 19.3 and Corollary 18.3. The desired conclusion now follows from Theorem 13.2. □

PROPOSITION 19.7. Let A be an α-Borel field of von Neumann algebras over H, let A_0 be a von Neumann algebra acting on a separable Hilbert space H_0, and let β be the coherence for $H(\cdot) \otimes H_0$ and W the linear isometry of $L^2(\mu;H(\cdot) \otimes H_0, \beta)$ onto $L^2(\mu;H,\alpha) \otimes H_0$ constructed in Lemma 7.4. Then $A(\cdot) \otimes A_0$ is a β-Borel field of von Neumann algebras over $H(\cdot) \otimes H_0$ and

$$W\left[\int_Z^\beta A(\xi) \otimes A_0 d\mu(\zeta)\right] W^{-1} = \left[\int_Z^\alpha A(\zeta) d\mu(\zeta)\right] \otimes A_0$$

Proof. This is an easy calculation using the definitions of β and W. □

THEOREM 19.8. Suppose that β is a coherence for a Borel field of Hilbert spaces K on Z and that A and B are α-Borel and β-Borel fields of von Neumann algebras over H and K, respectively. If $A(\zeta)$ and $B(\zeta)$ are spatially isomorphic [respectively, *-isomorphic] for μ-a.a. $\zeta \in Z$ then $\int_Z^\alpha A(\zeta) d\mu(\zeta)$ and $\int_Z^\beta B(\zeta) d\mu(\zeta)$ are spatially isomorphic [respectively, *-isomorphic].

Proof. This proof is completely analogous to that of Theorem 12.1. Considering first the assertion concerning spatial isomorphism, one may (by formula (*) of Section 18) assume that $A(\zeta)$ and $B(\zeta)$ are spatially isomorphic for each point ζ in Z, that H, K, α and β are all constant, and that $H = K$ and $\alpha = \beta$. This means that there is a Hilbert space H and that A and B are Borel maps from Z to $vN(H)$. Let U denote the group of unitary operators on H and put

$$W = \{(\zeta, U) \in Z \times U: UA(\zeta)U^* = B(\zeta)\}$$

If one knew that W were a Borel subset of $Z \times U$ then, just as before, one would be able to construct a unitary operator on $L^2(\mu;H)$ which implements a spatial isomorphism of $\int_Z^\oplus A(\zeta) d\mu(\zeta)$ and $\int_Z^\oplus B(\zeta) d\mu(\zeta)$. To show that W is actually a Borel set, select sequences of Borel maps (A_n) and (B_n) from Z to $L(H)$ such that for each point ζ in Z the sets $\{A_n(\zeta): n \in \mathbb{N}\}$ and $\{B_n(\zeta): n \in \mathbb{N}\}$ are *-algebras over $Q + iQ$ and are strongly dense in $A(\zeta)$ and $B(\zeta)$, respectively (such sequences exist by Theorem 17.1). Then a point (ζ, U) in $Z \times U$ will lie in W if and only if the two sets $\{UA_n(\zeta)U^*: n \in \mathbb{N}\}$ and $\{B_n(\zeta): n \in \mathbb{N}\}$ have the same strong closure. Now letting (x_p) be a dense sequence in H and putting

$$W(k,j,m,p) = \{(\zeta,U) \in Z \times U: \|[UA_j(\zeta)U^* - B_m(\zeta)]x_p\| < k^{-1}\}$$

for all $k,j,m,p \in \mathbb{N}$, it is not hard to see that

$$W = \cap_{k=1}^{\infty} \cap_{m=1}^{\infty} \cap_{r=1}^{\infty} \cup_{j=1}^{\infty} \cap_{p=1}^{r} [W(k,j,m,p) \cap W(k,m,j,p)]$$

and hence that W is Borel.

Turning now to the assertion concerning *-isomorphism, one may as well assume that $A(\zeta)$ and $B(\zeta)$ are *-isomorphic for each $\zeta \in Z$. But then $A(\zeta) \otimes C(\ell^2)$ and $B(\zeta) \otimes C(\ell^2)$ are spatially isomorphic for each $\zeta \in Z$ (Proposition A.3), and the assertion follows easily from Proposition 19.7 and what was just proven. □

COROLLARY 19.9. Let A be an α-Borel field of von Neumann algebras over H, and suppose that there is a von Neumann algebra A_0 which acts on a separable Hilbert space and is spatially isomorphic [respectively, *-isomorphic] to $A(\zeta)$ for μ-a.a. $\zeta \in Z$. Then $\int_Z^\alpha A(\zeta)d\mu(\zeta)$ and $A_0 \otimes M$ are spatially isomorphic [respectively, *-isomorphic].

Proof. If $B(\zeta) = A_0$ for each $\zeta \in Z$ then the von Neumann algebras $\int_Z^\alpha A(\zeta)d\mu(\zeta)$ and $\int_Z^\oplus B(\zeta)d\mu(\zeta)$ are spatially isomorphic [respectively, *-isomorphic] by the theorem, and $\int_Z^\oplus B(\zeta)d\mu(\zeta)$ and $A_0 \otimes M$ are in turn spatially isomorphic by Theorem 18.2 and Proposition 5.2. □

20. Examples

Only two examples of direct integral decompositions of von Neumann algebras acting on separable Hilbert spaces will be given. The first is a continuation of Example 10.3 and will show how one can, at least in principle, use structure theory to obtain the central decomposition of an algebra of

20. EXAMPLES

type I, and the second is a continuation of Example 10.9 and will consider how the decompositions of a von Neumann algebra with respect to two subalgebras of its center are related if one of the subalgebras in question contains the other.

EXAMPLE 20.1. In considering the direct integral decomposition of von Neumann algebras of type I one may, by virtue of the structure for such algebras (Theorem A.6), restrict one's attention to those of type I_n for some n. Accordingly, let M be an abelian von Neumann algebra acting on a separable Hilbert space and consider the corresponding type I_n algebra, $M \otimes L(\ell_n^2)$. From Example 10.3 one knows that M is spatially isomorphic to an algebra of the form $\int_Z^\alpha C(H(\zeta))d\mu(\zeta)$ for some standard Borel space Z, some finite Borel measure μ on Z, some Borel field of Hilbert spaces H on Z and some coherence α for H. But then one knows from Lemma 7.4 that there is a coherence β for the field of Hilbert spaces $H(\cdot) \otimes \ell_n^2$ with the property that $M \otimes C(\ell_n^2)$ and $M \otimes L(\ell_n^2)$ are spatially isomorphic to

$$\int_Z^\beta C(H(\zeta)) \otimes C(\ell_n^2)d\mu(\zeta)$$

and

$$\int_Z^\beta C(H(\zeta)) \otimes L(\ell_n^2)d\mu(\zeta)$$

respectively. This means, of course, that $\int_Z^\beta C(H(\zeta)) \otimes L(\ell_n^2)d\mu(\zeta)$ is the central decomposition of $M \otimes L(\ell_n^2)$. □

EXAMPLE 20.2. Retain the notation of Example 10.9 and, in addition, let A_0 be a von Neumann algebra acting on H with the property that its center contains M. Then from Theorem 19.4 one knows that there is an α-Borel field of von Neumann algebras A over H such that $UA_0U^{-1} = \int_X^\alpha A(\zeta)d\mu(\zeta)$ as well as a β-Borel field of von Neumann algebras B over K such that $VA_0V^{-1} = \int_Y^\beta B(\xi)d\nu(\xi)$.

Consider an α-Borel operator field A over H with the property that $\|A(\cdot)\|$ is a bounded function on Z. If v and w are any two α-Borel vector fields over H such that both $\|v(\cdot)\|$ and $\|w(\cdot)\|$ are bounded functions on Z then

$$\langle \int_X^\alpha A(\zeta)d\mu_\xi(\zeta) \int_X^\alpha v(\zeta)d\mu_\xi(\zeta), \int_X^\alpha w(\zeta)d\mu_\xi(\zeta) \rangle$$
$$= \int_X \langle A(\zeta)v(\zeta), w(\zeta)\rangle d\mu_\xi(\zeta)$$

is clearly a Borel function of ξ on Y. In view of Lemma 8.2 and the construction of the coherence γ, this means that $\xi \mapsto \int_X^\alpha A(\zeta)d\mu_\xi(\zeta)$ is a γ-Borel operator field over $\xi \mapsto L^2(\mu_\xi;H,\alpha)$. Now let (A_k) be an α-Borel generating sequence for A with $\|A_k(\zeta)\| \leq 1$ for all $\zeta \in X$ and all k. Then by Theorem 18.2 the operators $\int_X^\alpha \phi_j(\zeta)I_{H(\zeta)}d\mu_\xi(\zeta)$ together with the $\int_X^\alpha A_k(\zeta)d\mu_\xi(\zeta)$ generate the von Neumann algebra $\int_X^\alpha A(\zeta)d\mu_\xi(\zeta)$ for each $\xi \in Y$. These two remarks taken together show that $\xi \mapsto \int_X^\alpha A(\zeta)d\mu_\xi(\zeta)$ is a γ-Borel field of von Neumann algebras.

One can now form the von Neumann algebra $\int_Y^\gamma \int_X^\alpha A(\zeta)d\mu_\xi(\zeta)d\nu(\xi)$, and this algebra is easily seen to be a direct integral decomposition of A_0 with respect to N. In fact, the linear isometry WUV^{-1} of $\int_Y^\beta K(\xi)d\nu(\xi)$ onto $\int_Y^\gamma \int_X^\alpha H(\zeta)d\mu_\xi(\zeta)d\nu(\xi)$ induces a spatial isomorphism between $\int_Y^\beta B(\xi)d\nu(\xi)$ and $\int_Y^\gamma \int_X^\alpha A(\zeta)d\mu_\xi(\zeta)d\nu(\xi)$, and it follows from Theorem 19.5 that $T(\xi)$ even induces a spatial isomorphism between $B(\xi)$ and $\int_X^\alpha A(\zeta)d\mu_\xi(\zeta)$ for ν-a.a. $\xi \in Y$. Thus one may as well identify the two fields B and $\zeta \mapsto \int_X^\alpha A(\zeta)d\mu_\xi(\zeta)$.

In order to further illuminate Examples 10.9 and 15.4 one should examine the above calculation in the special case where A_0 and M coincide. Under this assumption one may clearly take $A(\zeta) = C(H(\zeta))$ for all $\zeta \in X$ and hence identify $B(\xi)$ with the algebra of diagonalizable operators on $L^2(\mu_\xi;H,\alpha)$ for ν-a.a. $\xi \in Y$. So in Example 10.9 [respectively, 15.4] the Hilbert space $\int_X^\alpha \pi(\zeta)d\mu_\xi(\zeta)$ [respectively, the representation $\int_X^\alpha \pi(\zeta)d\mu_\xi(\zeta)$] is the direct integral decomposition of $K(\xi)$ [respectively, $\rho(\xi)$] with respect to $B(\xi)$ for ν-a.a. $\xi \in Y$. □

HISTORICAL COMMENTS

The formulation of the direct integral theory of von Neumann algebras presented in this chapter differs somewhat from the usual one which is due to von Neumann [49] and is described in the books by Dixmier and Schwartz [10, 54]. This difference manifests itself in two ways: in the introduction of the Effros Borel structure, and in the use of the direct integral theory of representations of involutive Banach algebras. In the customary formulation of the theory a field of von Neumann algebras is defined to be Borel (or measurable) if it has an α-Borel generating sequence (in the terminology of Section 18) for a suitable coherence α. This definition was as utilitarian as it was inelegant, and the problem of constructing a Borel structure on vN(H) so as to make part (d) of Theorem 17.1 true presented itself naturally. This problem was solved by Effros [14] in 1965, who in fact proved all of Theorem 17.1. One should notice that Sections 18 and 19 do not depend on

20. EXAMPLES

this result of Effros, and that in fact the results contained therein as well as their proofs would make sense in the customary formulation of the theory of direct integrals of von Neumann algebras. The advantage of having the Effros Borel structure and Theorem 17.1 available will only become apparent in the following chapter.

What are here called Hausdorff metrics were introduced by Hausdorff in 1914 (cf. [26, p. 291]), and a number of results concerning them can be found in Kuratowski's books [32, 33]. Proposition 16.2 is due to Kuratowski [33, Section 38.I] and Lemma 16.3 to Effros [13, p. 930]. The proof of Theorem 17.1 given here was arrived at by combining Effros' original proof (in [14, Sections 2,3]) with some ideas of Maréchal (contained in [42, pp. 847-848]). In particular, Lemma 17.3 is taken from [42] and Lemmas 17.6 and 17.8 from [14].

Theorems 18.2, 19.1, 19.4 and 19.6 and Corollaries 18.3 and 19.2 are due to von Neumann [49, Sections 17-20], and all of the results of Sections 18 and 19 save Lemma 18.1 and Proposition 19.3 appear in Dixmier's book [10]. Proposition 19.3 is due to Ernest [18, Proposition 4]. Example 20.2 is due to Effros [16, Lemma 4.5] and has been discussed by Lance [34] and Nielsen [50, pp. 600-602].

Chapter 5
DIRECT INTEGRALS AND TYPES OF VON NEUMANN ALGEBRAS

21. The Statements of Two Theorems

The account of the Effros Borel structure and direct integral theory of von Neumann algebras contained in the last chapter is seriously incomplete in that it fails to consider the usual classification theory of von Neumann algebras (cf. Appendix A). The aim of the present chapter is to prove two theorems which will remedy this omission; roughly speaking, these theorems will say that both the Effros Borel structure and the direct integral theory are "as compatible as possible" with the classification theory. The present section simply contains the statements of these two theorems. The actual proofs of the theorems are interwoven with one another and depend on a number of technical lemmas; the theorems will be proven bit-by-bit throughout the following four sections.

In order to systematically discuss the classification of von Neumann algebras it will be convenient to let P_1 stand for "of type I", $P_{1,n}$ for "of type I_n", $n \in \mathbb{N} \cup \{\infty\}$, P_2 for "of type II", P_3 for "of type III", P_4 for "continuous", P_5 for "finite", P_6 for "semi-finite", P_7 for "properly infinite", P_8 for "abelian", and P_9 for "a factor". All but the last two of these properties are of course associated with central projections: for any von Neumann algebra A and any $k = 1, 2, \ldots, 7$ [respectively, $n = \infty, 1, 2, \ldots$], let $E_k(A)$ [respectively, $E_{1,n}(A)$] be zero if no direct summand of A is P_k [respectively, $P_{1,n}$] and otherwise let $E_k(A)$ [respectively, $E_{1,n}(A)$] be the largest central projection in A whose corresponding direct summand is P_k [respectively, $P_{1,n}$]. These projections satisfy the following relations:

$$E_1(A) + E_2(A) + E_3(A) = I$$
$$E_1(A) + E_4(A) = I$$
$$E_3(A) + E_6(A) = I \qquad\qquad\qquad (*)$$
$$E_5(A) + E_7(A) = I$$
$$E_1(A) = E_{1,\infty}(A) + E_{1,1}(A) + E_{1,2}(A) + \cdots$$

(cf. Appendix A). Finally, for any separable Hilbert space H put

$$vN_k(H) = \{A \in vN(H): A \text{ is } P_k\}, \quad 1 \leq k \leq 9$$
$$vN_1(H) = \{A \in vN(H): A \text{ is } P_{1,n}\}, \quad 1 \leq n \leq \infty$$

THEOREM 21.1. For any separable Hilbert space H

(i) the $vN_k(H)$, $1 \leq k \leq 9$, and the $vN_{1,n}(H)$, $1 \leq n \leq \infty$, are all Borel subsets of $vN(H)$, and

(ii) the E_k, $1 \leq k \leq 7$, and the $E_{1,n}$, $1 \leq n \leq \infty$, are all Borel functions from $vN(H)$ to $L(H)$.

THEOREM 21.2. If Z is a standard Borel space, if μ is a Borel measure on Z, if H is a Borel field of Hilbert spaces on Z, if α is a coherence for H, and if A is an α-Borel field of von Neumann algebras over H then

(i) $\int_Z^\alpha A(\zeta) d\mu(\zeta)$ is P_k, $1 \leq k \leq 7$ [respectively, $P_{1,n}$, $1 \leq n \leq \infty$] if and only if $A(\zeta)$ is P_k [respectively, $P_{1,n}$] for μ-a.a. $\zeta \in Z$, and

(ii) the $E_k \circ A$, $1 \leq k \leq 7$, and the $E_{1,n} \circ A$, $1 \leq n \leq \infty$, are α-Borel operator fields over H satisfying

$$\int_Z^\alpha E_k(A(\zeta)) d\mu(\zeta) = E_k\left(\int_Z^\alpha A(\zeta) d\mu(\zeta)\right)$$

and

$$\int_Z^\alpha E_{1,n}(A(\zeta)) d\mu(\zeta) = E_{1,n}\left(\int_Z^\alpha A(\zeta) d\mu(\zeta)\right)$$

The key to proving these two theorems turns out to be part (i) of Theorem 21.1. In Section 23 it will be shown that Theorem 21.2 (ii) is a corollary of Theorem 21.1 (ii), and in Section 22 that part (ii) of Theorem 21.1 is in turn a corollary of part (i). The proof of Theorem 21.1 (i) will

22. PARTIAL PROOF OF THEOREM 21.1

be begun in Section 22, where it will be shown that each of the sets in question is at least universally measurable. With the aid of this weak form of Theorem 21.1 (i) one can actually prove Theorem 21.2 (i), and all but the last paragraph of Section 23 is devoted to this derivation. Finally, one can use Theorem 21.2 (i) and two technical lemmas to complete the proof of Theorem 21.1 (i); Section 24 contains these technical lemmas and the proof itself occurs in Section 25.

22. Partial Proof of Theorem 21.1

The following notation will be fixed throughout this section: H will be a separable infinite-dimensional Hilbert space; W will be a linear isometry of $H \otimes H$ onto H and θ the map $A \mapsto W[A \otimes C(H)]W^{-1}$ of $vN(H)$ into itself; P_k, $P_{1,n}$, E_k, $E_{1,n}$, $vN_k(H)$, and $vN_{1,n}(H)$ will be as in Section 21; $U(H)$ will be the group of unitary operators on H; and for any nonempty subset S of $vN(H)$, $a(S)$ [respectively, $s(S)$] will denote the set consisting of all those algebras in $vN(H)$ which are *-isomorphic [respectively, spatially isomorphic] to some algebra in S. Thus θ is a one-one Borel mapping of $vN(H)$ into itself with the property that any two algebras A and B in $vN(H)$ are *-isomorphic if and only if $\theta(A)$ and $\theta(B)$ are spatially isomorphic (Proposition A.3); in particular,

$$a(S) = \theta^{-1}(s(\theta(S))) \tag{1}$$

for any subset S of $vN(H)$. The first two results of this section are immediate consequences of Theorem 17.1.

PROPOSITION 22.1. *Both of the sets $vN_8(H)$ and $vN_9(H)$ are Borel.*

LEMMA 22.2. *The map $(U,A) \mapsto UAU^*$ from $U(H) \times vN(H)$ onto $vN(H)$ is Borel.*

LEMMA 22.3. *The weak and strong operator topologies coincide on $U(H)$ and $U(H)$ is a topological group and a Polish space in this topology.*

Proof. The first two assertions are easy to verify. Let (x_j) be a dense sequence and (y_k) an orthonormal basis in H, and consider the sets

$$S(j,m,n) = \{A \in L(H): \sum_{k=1}^{m} |\langle Ax_j, y_k \rangle|^2 > \|x_j\|^2 - \tfrac{1}{n}\}$$

Then $\cap_j \cap_n \cup_m S(j,m,n)$ is a G_δ-subset of $L(H)$ in the weak operator topology and its intersection with the unit ball of $L(H)$ consists precisely of the isometric operators on H. Recalling that the unit ball of $L(H)$ is Polish in the weak operator topology and that an operator on H is unitary if and only if both it and its adjoint are isometric, the final assertion of the lemma follows immediately from Proposition 2.1. □

LEMMA 22.4. $a(\{A\})$ and $s(\{A\})$ are Borel subsets and $a(S)$ and $s(S)$ are analytic subsets of $vN(H)$ for any element A and any analytic subset S of $vN(H)$.

Proof. Fix an element A and an analytic subset S of $vN(H)$. Then $s(S)$ is just the image of $U(H) \times S$ by the map $(U,B) \mapsto UBU^*$ and so is analytic by Lemmas 2.5, 22.2 and 22.3. The set

$$U_0 = \{U \in U(H): UAU^* = A\}$$

is easily seen to be a closed subgroup of $U(H)$, and so by Lemma 22.3 and Corollary 3.3 there is a Borel subset, say U_1, of $U(H)$ meeting each left U_0-coset in exactly one point. But then $U \mapsto UAU^*$ is a one-one Borel mapping of U_1 onto $s(\{A\})$, and thus $s(\{A\})$ is a Borel set by Theorem 2.13. Finally, the assertions concerning $a(S)$ and $a(\{A\})$ follow from what has already been proven, from (1), and from Proposition 2.7. □

COROLLARY 22.5. The sets $vN_1(H)$ and $vN_{1,n}(H)$, $1 \leq n \leq \infty$, are all Borel.

Proof. Let $M_1, M_2, \ldots$ be mutually non-*-isomorphic abelian von Neumann algebras acting on H with the property that $a(\{M_k: k \in \mathbb{N}\}) = vN_8(H)$. Then

$$vN_{1,n}(H) = \cup_{k=1}^\infty a(\{W_n[M_k \otimes L(\ell_n^2)]W_n^{-1}\}), \quad 1 \leq n \leq \infty$$

and

$$vN_1(H) = \cup_{k=1}^\infty \hat{\theta}^{-1}(a(\{\hat{\theta}(M_k)\}))$$

where W_n is a fixed linear isometry of $H \otimes \ell_n^2$ onto H and where $\hat{\theta}(A) = [\theta(A')]'$ for all $A \in vN(H)$. Indeed, the first of these equalities is obvious, and the second is proven as follows. If A lies in $vN_1(H)$ then (by Theorem A.6) $A \otimes L(H)$ must be spatially isomorphic to a von Neumann algebra of the form $M \otimes L(H)$, where M is abelian, and hence *-isomorphic to $M_k \otimes L(H)$ for some k. But this means that $\hat{\theta}(A) \in a(\{\hat{\theta}(M_k)\})$. Conversely, say that $\hat{\theta}(A) \in a(\{\hat{\theta}(M_k)\})$ for some k and some $A \in vN(H)$. Then $A \otimes L(H)$ and $M_k \otimes L(H)$ are *-ismorphic, and thus $A \otimes L(H)$ and (by Theorem A.9) A itself are of type I. The corollary now follows from Lemma 22.4. □

22. PARTIAL PROOF OF THEOREM 21.1

COROLLARY 22.6. The set $vN_4(H)$ is coanalytic.

Proof. Let V be a linear isometry of $H \oplus H$ onto H and let ω be the map $(A,B) \mapsto V(A \oplus B)V^{-1}$ from $vN(H) \times vN(H)$ into $vN(H)$. Then ω is Borel by Theorem 17.1,

$$a(\omega(vN(H) \times vN_1(H))) = vN(H) - vN_4(H)$$

by Theorem A.8, and so the present corollary follows from the preceeding corollary and lemma. □

COROLLARY 22.7. The sets $E_{1,n}^{-1}(0) \cap vN_1(H)$, $1 \leq n \leq \infty$, are all Borel.

Proof. Let V be a linear isometry of $\oplus_{1 \leq j \leq \infty}(H \otimes \ell_j^2)$ onto H, let $M_1, M_2, \ldots$ be as in the proof of Corollary 22.5, and let K be the set of all functions from $\mathbb{N} \cup \{\infty\}$ to $\{0\} \cup \mathbb{N}$ with the product Borel structure. Put $K_n = \{a \in K: n \notin \text{range}(a)\}$ for $n \in \{0\} \cup \mathbb{N}$, put $M_{0,j} = C(H \otimes \ell_j^2)$ and $M_{k,j} = M_k \otimes L(\ell_j^2)$ for $k \in \mathbb{N}$ and $j \in \mathbb{N} \cup \{\infty\}$, and put $A(a) = V[\oplus_{1 \leq j \leq \infty} M_{a(j),j}]V^{-1}$ for $a \in K$. Then K_n is a Borel subset of K and A is a Borel mapping of K into $vN(H)$ with $a(A(K)) = vN_1(A)$, $a(A(K_n)) = E_{1,n}^{-1}(0) \cap vN_1(H)$, and

$$a(A(K - K_n)) = vN_1(H) - E_{1,n}^{-1}(0) \cap vN_1(H)$$

for $n \in \mathbb{N} \cup \{\infty\}$ (by Theorems 17.1 and A.6), and the corollary follows from Corollary 2.9 (the separation theorem), Lemma 22.4, and Corollary 22.5. □

PROPOSITION 22.8. The set $vN_5(H)$ is Borel.

Proof. To prove this result it is sufficient (by Corollary 2.9) to show that $vN_5(H)$ is both analytic and coanalytic. Let (A_n) be a sequence of Borel maps from $vN(H)$ to $L(H)$ with the property that for each A in $vN(H)$ the set $\{A_n(A): n \in \mathbb{N}\}$ is a *-algebra over $Q + iQ$ which is strongly dense in A. Now the complement of $vN_5(H)$ in $vN(H)$ is just the image of the set

$$\{(A,V) \in vN(H) \times L(H): V \in A \text{ and } V^*V \neq VV^* = I\}$$

under the projection of $vN(H) \times L(H)$ onto its first coordinate, and this set is Borel by Theorem 17.1 (e) and the fact that a point (A,V) in $vN(H) \times L(H)$ satisfies the condition $V \in A$ if and only if $VA_n(A') = A_n(A')V$ for all n. This shows that $vN_5(H)$ is at least coanalytic.

For each pair of vectors x,y in H let $E_{x,y}$ denote the rank-one operator $z \mapsto \langle z,y \rangle x$ on H. Then for each linear functional ϕ in $L(H)_*^+$ there is (by the Riesz representation theorem) a unique positive operator T_ϕ on H satisfying $\langle T_\phi x,y \rangle = \phi(E_{x,y})$, $x,y \in H$. The set

$$W = \bigcap_{m=1}^{\infty} \bigcap_{n=1}^{\infty} \bigcup_{k=1}^{\infty} \bigcup_{j=1}^{\infty} \{(\phi,A) \in L(H)_*^+ \times vH(H): \phi(A_m(A)A_n(A)) = \phi(A_n(A)A_m(A)) \text{ and } \|A_k(A')T_\phi x_j - x_m\| < \tfrac{1}{n}\}$$

where (x_j) is some fixed dense sequence in H, is Borel by Theorem 17.1 (e) and the fact that multiplication is jointly Borel on $L(H)$. A point (ϕ,A) in $L(H)_*^+ \times vN(H)$ clearly lies in W if and only if the restriction of ϕ to A is a trace and if $A'T_\phi H$ is dense in H. So to complete the proof of the proposition it is sufficient (by Theorem A.11) to show that the restriction of a linear functional ϕ in $L(H)_*^+$ to an A in $vN(H)$ is faithful if and only if $A'T_\phi H$ is dense in H.

Accordingly, let (ϕ,A) be a point in $L(H)_*^+ \times vN(H)$ and consider the projection E of H onto $(A'T_\phi H)^\perp$. Then E belongs to A. If $E \neq 0$ and if (y_j) is an orthonormal basis for EH then $E = \sum_j E_{y_j,y_j}$ and

$$\phi(E) = \sum_j \phi(E_{y_j,y_j}) = \sum_j \langle T_\phi y_j, y_j \rangle = 0$$

Conversely, if the restriction of ϕ to A^+ is not faithful there will be (by spectral theory) a nonzero projection F in A with $\phi(F) = 0$. Then for all vectors x in FH, all vectors y in H, and all operators A in A' one has

$$\begin{aligned}|\langle x, AT_\phi y \rangle|^2 &= |\langle T_\phi FA^*x, y \rangle|^2 \\ &= |\phi(FA^*E_{x,y})|^2 \\ &= |\langle A^*E_{x,y}, F \rangle_\phi|^2 \\ &\leq \langle A^*E_{x,y}, A^*E_{x,y} \rangle_\phi \langle F,F \rangle_\phi \\ &= 0\end{aligned}$$

where $\langle \cdot, \cdot \rangle_\phi$ denotes the semi-inner product ϕ induces on $L(H)$. This shows that $F \leq E$, and hence that $E \neq 0$. □

COROLLARY 22.9. *The set* $vN_6(H)$ *is analytic.*

Proof. One has $vN_6(H) = a(vN_5(H)')$ by Theorem A.7, and therefore $vN_6(H)$ is analytic by Theorem 17.1 (e), Lemma 22.4, and Proposition 22.8. □

22. PARTIAL PROOF OF THEOREM 21.1

COROLLARY 22.10. The set $vN_2(H)$ is universally measurable.

Proof. This is an immediate consequence of Corollaries 22.6 and 22.9. □

COROLLARY 22.11. The set $vN_7(H)$ is coanalytic.

Proof. If ω is the Borel map from $vN(H) \times vN(H)$ into $vN(H)$ constructed in the proof of Corollary 22.6 then

$$vN(H) - vN_7(H) = a(\omega(vN(H) \times vN_5(H)))$$

by the definition of a von Neumann algebra being properly infinite, and so the corollary follows from Proposition 22.8 and Lemma 22.4. □

The rest of this section is concerned with the functions E_k and $E_{1,n}$ and the set $vN_3(H)$. The study of these functions and this set necessarily involves the reduction of algebras in $vN(H)$, and the analysis of this process and the algebras which result from it will be carried out by means of a construction which will now be described. The point of this construction is to systematically replace all such algebras by *-isomorphic ones acting on the fixed Hilbert space H.

Put $J(n) = \{1,2,\ldots,n\}$ for $n \in \mathbb{N}$, put $J(\infty) = \mathbb{N}$, and let η_n be a bijection of $J(n) \times J(\infty)$ onto $J(\infty)$ for $n \in \mathbb{N} \cup \{\infty\}$. Let $e_1, e_2, \ldots$ be an orthonormal basis in H and let $P(H)$ denote the set of all projections on H. For each projection E in $P(H)$ one can apply the Gram-Schmidt process to the sequence (Ee_j) to obtain a sequence of vectors $(f_{E,j})$ in EH such that

(a) if $E \neq 0$ the $f_{E,j}$, $j \in J(\dim EH)$, form an orthonormal basis in EH and the $f_{E,j}$, $j > \dim EH$, are all zero, and

(b) $E \mapsto f_{E,j}$ is a Borel function from $P(H)$ to H for each j (cf. Section 8).

This shows, in particular, that the sets

$$P_n(H) = \{E \in P(H): \dim EH = n\}, \quad n \in \mathbb{N} \cup \{\infty\}$$

are all Borel. For each $n \in \mathbb{N} \cup \{\infty\}$ and each $E \in P_n(H)$ let W_E be the unique linear isometry of $EH \otimes H$ onto H which carries $f_{E,j} \otimes e_k$ onto $e_{\eta_n(j,k)}$, $j \in J(n)$ and $k \in \mathbb{N}$. Finally, let θ_0 be the map $(A,E) \mapsto W_E[A_E \otimes C(H)]W_E^{-1}$ from

$$\{(A,E) \in vN(H) \times P(H): 0 \neq E \in A\}$$

into $vN(H)$.

LEMMA 22.12. The domain of θ_0 is a Borel subset of $vN(H) \times P(H)$ and θ_0 itself is a Borel map.

Proof. A point (A,E) in $vN(H) \times P(H)$ satisfies $E \in A$ if and only if it satisfies $EA_n(A') = A_n(A')E$ for all n (here the A_n are as in the proof of Proposition 22.8), and so the first assertion follows from Theorem 17.1 (e). Now fix an element n in $\mathbb{N} \cup \{\infty\}$ and elements j, k, and m in $\mathbb{N}$, and put $\eta_n^{-1}(j) = (a,b)$ and $\eta_n^{-1}(k) = (c,d)$. Then for any pair (A,E) in the domain of θ_0 with $E \in P_n(H)$ one has

$$\langle W_E[A_m(A)_E \otimes I]W_E^{-1} e_j, e_k \rangle$$
$$= \langle [A_m(A)_E \otimes I] f_{E,a} \otimes e_b, f_{E,c} \otimes e_d \rangle$$
$$= \delta_{b,d} \langle A_m(A) f_{E,a}, f_{E,c} \rangle$$

and so the second assertion follows from Theorem 17.1 (d). □

COROLLARY 22.13. The set $vN_3(H)$ is coanalytic.

Proof. The complement of $vN_3(H)$ in $vN(H)$ is just the image of the set $\theta_0^{-1}(vN_5(H))$ under the projection of $vN(H) \times P(H)$ onto its first coordinate, and this image is analytic by Proposition 22.8 and Lemma 22.12. □

COROLLARY 22.14. If all of the $vN_k(H)$, $1 \leq k \leq 7$, are Borel sets then all of the E_k, $1 \leq k \leq 7$, and the $E_{1,n}$, $1 \leq n \leq \infty$, will be Borel functions.

Proof. Let j be one of 1, 3, or 5 and let $k = 4$ if $j = 1$, let $k = 6$ if $j = 3$, and let $k = 7$ if $j = 5$. Then $E_j^{-1}(I) = vN_j(H)$ and $E_j^{-1}(0) = vN_k(H)$ by (*) of Section 21, and the graph of the restriction of E_j to $vN(H) - vN_j(H) \cup vN_k(H)$ is $\theta_0^{-1}(vN_j(H)) \cap \theta_I^{-1}(vN_k(H))$; here θ_I is the map $(A,E) \mapsto \theta_0(A, I - E)$ of $\{(A,E) \in vN(H) \times P(H): I \neq E \in A\}$ into $vN(H)$. Thus E_1, E_3 and E_5 are Borel functions by Lemma 22.12 and Corollary 2.11, and hence each of the E_k, $1 \leq k \leq 7$, is a Borel function by (*) of Section 21.

Now fix an element n of $\mathbb{N} \cup \{\infty\}$. The first step in showing that $E_{1,n}$ is a Borel function is to show that its restriction, say G_n, to $vN_1(H)$ is Borel. Now $G_n^{-1}(0)$ and $G_n^{-1}(I)$ are both Borel subsets of $vN(H)$ by Corollary 22.7, and hence $\theta_0^{-1}(G_n^{-1}(0)) \cap \theta_I^{-1}(G_n^{-1}(I))$ is also a Borel subset of $vN(H)$ by Lemma 22.12. But this latter set is just the graph of the restriction

23. PROOF OF THEOREM 21.2

of G_n to $vN_1(H) - G_n^{-1}(\{0,I\})$, and so G_n is Borel by Corollary 2.11. Next, notice that $vN(H)$ is the disjoint union of the Borel sets $E_1^{-1}(P_m(H))$, $m \in \mathbb{N} \cup \{\infty\}$, and that $E_{1,n} = 0$ on $E_1^{-1}(P_0(H))$, and hence that it is sufficient to show that the restriction of $E_{1,n}$ to $E_1^{-1}(P_m(H))$ is Borel for any $m \in \mathbb{N} \cup \{\infty\}$. Fix such an m. Then there will be Borel functions $a_{p,j}$ on $P_m(H)$, $p \in J(\infty)$ and $j \in J(m)$, satisfying

$$Ee_p = \sum_{j=1}^{m} a_{p,j}(E)f_{E,j}, \qquad E \in P_m(H) \text{ and } p \in J(m)$$

(cf. Section 8). So if A lies in $E^{-1}(P_m(H))$ and if p and q lie in $\mathbb{N}$ then

$$\langle E_{1,n}(A)e_p, e_q \rangle$$
$$= \langle E_{1,n}(A)E_1(A)e_p, E_1(A)e_q \rangle$$
$$= \sum_{j,k=1}^{m} a_{p,j}(E_1(A))\overline{a_{q,k}(E_1(A))}\langle E_{1,n}(A)f_{E_1(A),j}, f_{E_1(A),k}\rangle$$

and, moreover, if j and k lie in $J(m)$ then

$$\langle E_{1,n}(A)f_{E_1(A),j}, f_{E_1(A),k}\rangle$$
$$= \langle W_{E_1(A)}[E_{1,n}(A)_{E_1(A)} \otimes I]W_{E_1(A)}^{-1} e_{\eta_m(j,1)}, e_{\eta_m(k,1)}\rangle$$
$$= \langle G_n(\theta_0(A, E_1(A)))e_{\eta_m(j,1)}, e_{\eta_m(k,1)}\rangle$$

From Lemma 22.12 and this calculation it should now be clear that the restriction of $E_{1,n}$ to $E_1^{-1}(P_m(H))$ is in fact a Borel function. □

23. Proof of Theorem 21.2

In proving Theorem 21.2 one may as well assume that the field H and the coherence α are constant (cf. Section 18). Thus A is a Borel map from Z to $vN(H)$ for some separable Hilbert space H. Let $P(H)$ be the set of all projections on H and let (A_j) be a fixed sequence of Borel maps from $vN(H)$ to $L(H)$ with the property that the set $\{A_j(A): j \in \mathbb{N}\}$ is self-adjoint and strongly dense in A for each $A \in vN(H)$ (cf. Theorem 17.1). The type of $\int_Z^\alpha A(\zeta)d\mu(\zeta)$, which is what Theorem 21.2 (i) is concerned with, is of course defined through central supports of projections in and the reduction of this algebra, and it is therefore appropriate to begin with some lemmas on projections, central supports, and reduction. But first a definition. For any pair (A,E) in $vN(H) \times P(H)$ let $P(A,E)$ denote the projection of H onto the closure of AEH. The interest in this function P resides in the fact that if E lies in A then $P(A,E)$ is the central support of E in A.

LEMMA 23.1. P is a Borel map from $vN(H) \times P(H)$ to $P(H)$.

Proof. Let (x_k) be a dense sequence in H. Then for any point (A,E) in $vN(H) \times P(H)$ the vectors of the form $A_j(A)Ex_k$ will be dense in the range of $P(A,E)$ and therefore

$$<P(A,E)y,y> = \|P(A,E)y\|^2$$
$$= \sup_{j,k} |<y,[A_j(A)Ex_k]>|^2$$

for any $y \in H$; here $[\cdot]$ is as in Section 8. This clearly proves that P is a Borel function. □

COROLLARY 23.2. Suppose that E is a Borel function from Z to $P(H)$ with the property that $E(\zeta) \in A(\zeta)$ for each $\zeta \in Z$, and let $F(\zeta)$ be the central support of $E(\zeta)$ in $A(\zeta)$, $\zeta \in Z$. Then F is Borel and $F(\mu)$ is the central support of $E(\mu)$ in $\int_Z^\oplus A(\zeta)d\mu(\zeta)$. In particular, the central support of $E(\mu)$ in $\int_Z^\oplus A(\zeta)d\mu(\zeta)$ is the identity if and only if the central support of $E(\zeta)$ in $A(\zeta)$ is the identity for μ-a.a. $\zeta \in Z$.

Proof. That F is a Borel function follows from Lemma 23.1 and the fact that $F(\zeta) = P(A(\zeta),E(\zeta))$, $\zeta \in Z$. Recall that an arbitrary projection in the center of $\int_Z^\oplus A(\zeta)d\mu(\zeta)$ has the form $G(\mu)$ for some Borel function G from Z to $P(H)$ with $G(\zeta) \in A(\zeta) \cap A(\zeta)'$ for all $\zeta \in Z$ (Theorem 19.1). Now such a projection is larger than $E(\mu)$ if and only if $G(\zeta) \geq E(\zeta)$ for μ-a.a. $\zeta \in Z$, hence if and only if $G(\zeta) \geq F(\zeta)$ for μ-a.a. $\zeta \in Z$, and, finally, if and only if $G(\mu) \geq F(\mu)$. This proves the second assertion, and the last one is now obvious. □

LEMMA 23.3. Let E be a Borel function from Z to $P(H)$ with the property that $E(\zeta) \in A(\zeta)$ for each $\zeta \in Z$, put $Y = \{\zeta \in Z: E(\zeta) \neq 0\}$ and $K(\zeta) = E(\zeta)H$ for $\zeta \in Y$, and assume that $\mu(Y) > 0$. Then K is a Borel field of Hilbert spaces on Y and there is a coherence β for K and a linear isometry W of $E(\mu)L^2(\mu;H)$ onto $L^2(\nu;K,\beta)$, where ν is the restriction of μ to the Borel subsets of Y, such that

23. PROOF OF THEOREM 21.2

(i) $\zeta \mapsto A(\zeta)_{E(\zeta)}$ is a β-Borel field of von Neumann algebras over K, and

(ii) $\left[\int_Z^\oplus A(\zeta)d\mu(\zeta)\right]_{E(\mu)} = W^{-1}\left[\int_Y^\beta A(\zeta)_{E(\zeta)}d\nu(\zeta)\right]W$.

Proof. It follows easily from Section 8 that K is Borel and that there is a coherence β for K such that the restrictions to Y of the vector fields $\zeta \mapsto E(\zeta)x$, $x \in H$, are β-Borel and, moreover, that if v is a function in $L^2(\mu;H)$ satisfying $E(\zeta)v(\zeta) = v(\zeta)$, $\zeta \in Z$, then the restriction v_Y of v to Y is a β-Borel vector field over K with

$$\left\|\int_Y^\beta v_Y(\zeta)d\nu(\zeta)\right\| = \left\|\int_Z^\oplus v(\zeta)d\mu(\zeta)\right\|$$

Conversely, given a function w in $L^2(\nu;K,\beta)$, the map v from Z to H which equals w on Y and zero off Y clearly satisfies $v \in L^2(\mu;H)$ and $v_Y = w$. This shows that the map $v \mapsto v_Y$ induces a linear isometry W of the range of $E(\mu)$ onto $L^2(\mu;K,\beta)$.

Given a Borel map A from Z to $L(H)$, let A_E denote the operator field $\zeta \mapsto A(\zeta)_{E(\zeta)}$ over K. Then it is not hard to show that such an A_E is β-Borel (use Lemma 8.2) and that

$$\left[\int_Z^\oplus A(\zeta)d\mu(\zeta)\right]_{E(\mu)} = W^{-1}\left[\int_Y^\beta A(\zeta)_{E(\zeta)}d\nu(\zeta)\right]W$$

if A is μ-essentially bounded, that $\zeta \mapsto A(\zeta)_{E(\zeta)}$ is a β-Borel field of von Neumann algebras (use Proposition A.5), and hence that

$$\left[\int_Z^\oplus A(\zeta)d\mu(\zeta)\right]_{E(\mu)} \subset W^{-1}\left[\int_Y^\beta A(\zeta)_{E(\zeta)}d\nu(\zeta)\right]W$$

To complete the proof of the lemma it is only necessary to notice that if B is a ν-essentially bounded β-Borel operator field over K with $B(\zeta) \in A(\zeta)_{E(\zeta)}$ for all $\zeta \in Y$ then the map A from Z to $L(H)$ defined by putting $A(\zeta) = B(\zeta)E(\zeta)$ for $\zeta \in Y$ and $A(\zeta) = 0$ for $\zeta \notin Y$ is μ-essentially bounded and Borel and satisfies $A(\zeta) \in A(\zeta)$ for all $\zeta \in Z$ and

$$\left[\int_Z^\oplus A(\zeta)d\mu(\zeta)\right]_{E(\mu)} = W^{-1}\left[\int_Y^\beta B(\zeta)d\nu(\zeta)\right]W$$

COROLLARY 23.4. If E is a Borel map from Z to $P(H)$ with the property that $E(\zeta) \in A(\zeta)$ for each $\zeta \in Z$ then $E(\mu)$ is an abelian projection in $\int_Z^\oplus A(\zeta)d\mu(\zeta)$ if and only if $E(\zeta)$ is an abelian projection in $A(\zeta)$ for μ-a.a. $\zeta \in Z$.

Proof. This follows immediately from the lemma and Corollary 18.3, Theorem 19.1, and Proposition A.5. □

LEMMA 23.5. *If E is a Borel map from Z to $P(H)$ with the property that $E(\zeta) \in A(\zeta)$ for each $\zeta \in Z$ then $E(\mu)$ is a finite projection in $\int_Z^\oplus A(\zeta)d\mu(\zeta)$ if and only if $E(\zeta)$ is a finite projection in $A(\zeta)$ for μ-a.a. $\zeta \in Z$.*

Proof. Suppose that $E(\mu)$ is an infinite projection in $\int_Z^\oplus A(\zeta)d\mu(\zeta)$. Then by definition there is a Borel map V from Z to $L(H)$ with the property that $V(\zeta)$ is a partial isometry in $A(\zeta)$ for each $\zeta \in Z$ and that

$$V(\mu)V(\mu)^* < V(\mu)^*V(\mu) = E(\mu)$$

But then $V(\zeta)V(\zeta)^* \leq V(\zeta)^*V(\zeta) = E(\zeta)$ for μ-a.a. $\zeta \in Z$ and $V(\zeta)V(\zeta)^* < V(\zeta)^*V(\zeta)$ on a Borel set of positive μ-measure, and so it is not the case that $E(\zeta)$ is a finite projection in $A(\zeta)$ for μ-a.a. $\zeta \in Z$.

Now suppose that, conversely, the set

$$Y = \{\zeta \in Z : E(\zeta) \text{ is an infinite projection in } A(\zeta)\}$$

fails to be μ-null. Letting θ_0 be as in Section 22, the map $\zeta \mapsto \theta_0(A(\zeta), E(\zeta))$ from the Borel subset $\{\zeta \in Z : E(\zeta) \neq 0\}$ of Z to $vN(H)$ is Borel, and hence the inverse image of $vN_5(H)$ is a Borel subset of Z by Proposition 22.8. But this inverse image is just $Z - Y$, and therefore Y is a Borel subset of Z with $\mu(Y) > 0$. Now consider the set

$$W = \{(\zeta, V) \in Y \times L(H) : V \in A(\zeta) \text{ and } V^*V < VV^* = E(\zeta)\}$$

This is a Borel subset of $Y \times L(H)$ (cf. the proof of Lemma 22.8) whose image under the projection of $Y \times L(H)$ onto its first coordinate is all of Y. It then follows from Theorem 4.3 that there is a Borel map V from Z to $L(H)$ satisfying $V(\zeta) = E(\zeta)$ for $\zeta \notin Y$ and $(\zeta, V(\zeta)) \in W$ for μ-a.a. $\zeta \in Y$. But then $V(\mu)$ is a partially isometric operator in $\int_Z^\oplus A(\zeta)d\mu(\zeta)$ satisfying

$$V(\mu)^*V(\mu) < V(\mu)V(\mu)^* = E(\mu)$$

and therefore $E(\mu)$ is in fact an infinite projection in $\int_Z^\oplus A(\zeta)d\mu(\zeta)$. □

Now that these technical preliminaries have been dispensed with, one can begin the actual proof of Theorem 21.2 (i).

23. PROOF OF THEOREM 21.2

Case $k = 5$. This is just a special case of Lemma 23.5.

Case $k = 6$. If $\int_Z^\oplus A(\zeta)d\mu(\zeta)$ is semi-finite then it is clear from Corollary 23.2 and Lemma 23.5 that $A(\zeta)$ is semi-finite for μ-a.a. $\zeta \in Z$. In proving the converse, one may as well assume that each $A(\zeta)$ is semi-finite. Consider the set

$$W = \{(\zeta,E) \in Z \times P(H): \text{ E is a finite projection in } A(\zeta) \text{ whose central support in } A(\zeta) \text{ is I}\}$$

The image of this set under the projection of $Z \times P(H)$ onto its first coordinate is all of Z and, moreover, this set is a Borel subset of $Z \times P(H)$ by virtue of the observation that a nonzero projection E in $A(\zeta)$ is finite in $A(\zeta)$ if and only if $\theta_0(A(\zeta),E) \in vN_5(H)$ together with Lemmas 22.12 and 23.1 and Proposition 22.8. Thus one can use Theorem 4.3 to find a Borel function E from Z to $P(H)$ such that $E(\zeta) \in A(\zeta)$ for all $\zeta \in Z$ and $(\zeta,E(\zeta)) \in W$ for μ-a.a. $\zeta \in Z$, and then, in turn, Corollary 23.2 and Lemma 23.5 to deduce that $\int_Z^\oplus A(\zeta)d\mu(\zeta)$ is semi-finite.

Case $k = 3$. If $\int_Z^\oplus A(\zeta)d\mu(\zeta)$ is not of type III then it must contain a nonzero finite projection, and this projection will necessarily be of the form $E(\mu)$ for some Borel map E from Z to $P(H)$ with $E(\zeta) \in A(\zeta)$ for all $\zeta \in Z$. But then (by Lemma 23.5) it will not be the case that $A(\zeta)$ is of type III for μ-a.a. $\zeta \in Z$. To prove the converse, first notice that the set

$$Y = \{\zeta \in Z: A(\zeta) \text{ is not of type III}\}$$

is analytic by Corollary 22.13, and hence μ-measurable by Theorem 4.1. This means there are Borel sets Y_1 and Y_2 in Z with $Y_1 \subset Y \subset Y_2$ and $\mu(Y_2 - Y_1) = 0$. Next, notice that

$$W = \{(\zeta,E) \in Y_1 \times P(H): \text{ E is a nonzero finite projection in } A(\zeta)\}$$

is a Borel subset of $Y_1 \times P(H)$ (cf. the proof of the case $k = 6$) whose image under the projection of $Y_1 \times P(H)$ onto its first coordinate is all of Y_1. Thus by Theorem 4.3 there is a μ-null Borel set N in Y_1 and a Borel map E from $Y_1 - N$ to $P(H)$ with $(\zeta,E(\zeta)) \in W$ for all $\zeta \in Y_1 - N$. Now putting $E(\zeta) = 0$ for $\zeta \in Z - (Y_1 - N)$, one obtains a Borel map E from Z to $P(H)$ with $E(\zeta) \in A(\zeta)$ for all $\zeta \in Z$ and with the property that $E(\mu)$ is a finite projection in $\int_Z^\oplus A(\zeta)d\mu(\zeta)$ (Lemma 23.5). So if $\int_Z^\oplus A(\zeta)d\mu(\zeta)$ is of type III then $E(\mu) = 0$, hence $\mu(Y_1) = \mu(Y_2) = 0$, and so Y is μ-null.

Case $k = 7$. This is similar to the case $k = 3$, the only changes being that one must use Corollary 22.11 instead of Corollary 22.13 and that the projections in question must now be central. But this latter change will not cause any complications in view of Theorems 17.1 (e) and 19.1.

Case $k = 1$. It is clear from Corollaries 23.2 and 23.4 that if $\int_Z^\oplus A(\zeta)d\mu(\zeta)$ is of type I then so is $A(\zeta)$ for μ-a.a. $\zeta \in Z$. In proving the converse one may as well assume that each $A(\zeta)$ is of type I. If

$$W = \{(\zeta, E) \in Z \times P(H): \text{ E is an abelian projection in } A(\zeta) \text{ whose central support is I}\}$$

was known to be a Borel subset of $Z \times P(H)$ then it would follow from Corollaries 23.2 and 23.4 that $\int_Z^\oplus A(\zeta)d\mu(\zeta)$ was of type I (cf. the proof of the case $k = 6$). Now a point (ζ, E) in $Z \times P(H)$ lies in W if and only if $E \in A(\zeta)$, $P(A(\zeta), E) = I$, and $\theta_0(A(\zeta), E) \in vN_8(H)$ (here θ_0 is the Borel function introduced in Section 22), and so W is Borel by Lemmas 22.12 and 23.1 and Proposition 22.1.

Case $k = 4$. If $\int_Z^\oplus A(\zeta)d\mu(\zeta)$ fails to be continuous then (by Theorem A.8) it will contain a nonzero central projection whose corresponding direct summand is of type I. This means (by Theorem 19.1) that there is a Borel map E from Z to $P(H)$ with $E(\zeta) \in A(\zeta) \cap A(\zeta)'$ for each $\zeta \in Z$, with $\mu(Y) > 0$, where $Y = \{\zeta \in Z: E(\zeta) \neq 0\}$, and with $\left[\int_Z^\oplus A(\zeta)d\mu(\zeta)\right]_{E(\mu)}$ of type I. But then $A(\zeta)_{E(\zeta)}$ will be of type I for μ-a.a. $\zeta \in Y$ by Lemma 23.3 and the case $k = 1$. Now suppose that, conversely, it is not the case that $A(\zeta)$ is continuous for μ-a.a. $\zeta \in Z$. Then the set $\{\zeta \in Z: A(\zeta) \text{ is not continuous}\}$ is μ-measurable by Corollary 22.6 and Theorem 4.1, and therefore contains a Borel set Y with $\mu(Y) > 0$. If

$$W = \{(\zeta, E) \in Y \times P(H): 0 \neq E \in A(\zeta) \cap A(\zeta)' \text{ and } A(\zeta)_E \text{ is of type I}\}$$

was known to be a Borel subset of $Y \times P(H)$ then it would follow from Lemma 23.3, Theorem A.8, and the case $k = 1$ that $\int_Z^\oplus A(\zeta)d\mu(\zeta)$ was not continuous (cf. the proof of the case $k = 6$). But a by-now standard argument using the function θ_0 introduced in Section 22 will show that W is in fact a Borel set.

Case $k = 2$. This is an immediate consequence of the cases $k = 4$ and $k = 6$.

Case $P_{1,n}$. Let $J(n)$ be as in Section 22. If $\int_Z^\oplus A(\zeta)d\mu(\zeta)$ is of type I_n then one can find a family $(E_j)_{j \in J(n)}$ of Borel maps from Z to $P(H)$ such that

24. SOME TECHNICAL LEMMAS

$E_j(\zeta) \in A(\zeta)$ for each j and each ζ and such that the $E_j(\mu)$ are mutually orthogonal and equivalent abelian projections in $\int_Z^\oplus A(\zeta)d\mu(\zeta)$ whose sum is the identity. But then (by a simple argument and Corollary 23.4) it must be the case that for μ-a.a. $\zeta \in Z$, the $E_j(\zeta)$ are mutually orthogonal and equivalent abelian projections in $A(\zeta)$ whose sum is the identity, i.e., that $A(\zeta)$ is of type I_n. In proving the converse one may as well assume that each $A(\zeta)$ is of type I_n. Then a by-now standard argument involving the set

$$\{(\zeta, (E_j)_{j \in J(n)}) \in Z \times P(H)^{J(n)}: \text{ the } E_j \text{ are mutually orthogonal and}$$
$$\text{equivalent abelian projections in } A(\zeta) \text{ whose sum is the identity}\}$$

where $P(H)^{J(n)}$ is the set of all functions from $J(n)$ to $P(H)$ with the product Borel structure, will show that $\int_Z^\oplus A(\zeta)d\mu(\zeta)$ is of type I_n.

Under the assumption that Theorem 21.1 (i) has been established it is now an easy matter to prove Theorem 21.2 (ii). Recall from Corollary 22.14 that the assumption implies that the $E_k(A(\cdot))$, $1 \leq k \leq 7$, and the $E_{1,n}(A(\cdot))$, $1 \leq n \leq \infty$, are Borel functions from Z to $P(H)$. To prove the formulae concerning the E_k it is enough, in view of formulae (*) of Section 21, to show that $(E_k \circ A)(\mu) \leq E_k(\int_Z^\oplus A(\zeta)d\mu(\zeta))$ for $1 \leq k \leq 7$. Now for a fixed k this inequality is obviously true if the set $\{\zeta \in Z: E_k(A(\zeta)) \neq 0\}$ is μ-null, and otherwise is a consequence of Lemma 23.3 and what was just proven (i.e., Theorem 21.2 (i)). Finally, a similar argument will serve to prove the formulae involving the $E_{1,n}$.

24. Some Technical Lemmas

As has already been mentioned, the object of the present section is to prove two rather technical lemmas (Lemmas 24.4 and 24.7) which will be essential in completing the proof of Theorem 21.1 (i). Because of the nature of these lemmas the reader would be well-advised to omit their proofs until after Section 25 has been read. The point of the first of these two lemmas is to enable one to deduce the analyticity of certain sets of von Neumann algebras from that of corresponding sets of factors. The other lemma simply asserts that the set of factors of type III is a Borel set, and its proof is quickly reduced to that of showing that the same set is analytic. This innocuous-sounding task turns out to be surprisingly difficult. Roughly speaking, the reason for this difficulty is as follows. To show that a set is analytic one has to be able to describe it as the range of some Borel map. If the set in question is the set of factors of type III acting on some separable Hilbert space then this effectively means that one has to be able to distinguish the factors of type III from amongst all of the factors by countable many "Borel" conditions, and it is precisely here that the difficulty resides. In fact, to show by "classical" techniques (i.e., ones independent of the Takesaki-Tomita theory) that a factor is of type III one must prove either that all of its nonzero projections are infinite (an uncountable number of conditions) or else that it has no semi-finite faithful normal trace (which really describes the complementary set and hence proves coanalyticity).

LEMMA 24.1. Suppose that Z and Y are two standard Borel spaces, that ω is a Borel measure on Z, that μ is a Borel field of Hilbert spaces on Z, and that α is a coherence for H. Put $\nu = \omega_*(\mu)$ and let $\xi \mapsto \mu_\xi$ be the Borel map from Y into $M(Z)$ obtained by applying Theorem 4.5 to Z,Y,ω and μ. Then $\xi \mapsto L^2(\mu_\xi;H,\alpha)$ is a Borel field of Hilbert spaces on Y and there is a coherence β for this field satisfying

(i) if v [respectively, A] is an α-Borel vector field [respectively, operator field] over H such that $\|v(\cdot)\|$ [respectively $\|A(\cdot)\|$] is a bounded function on Z then $\xi \mapsto \int_Z^\alpha v(\zeta)d\mu_\xi(\zeta)$ [respectively, $\xi \mapsto \int_Z^\alpha A(\zeta)d\mu_\xi(\zeta)$] is a β-Borel vector field [respectively, operator field] over $\xi \mapsto L^2(\mu_\xi;H,\alpha)$,

(ii) there is a linear isometry W of $\int_Z^\alpha H(\zeta)d\mu(\alpha)$ onto $\int_Y^\beta \int_Z^\alpha H(\zeta)d\mu_\xi(\zeta)d\nu(\xi)$ such that if v and A are as in (i) then

$$W \int_Z^\alpha v(\zeta)d\mu(\zeta) = \int_Y^\beta \int_Z^\alpha v(\zeta)d\mu_\xi(\zeta)d\nu(\xi)$$

and

$$W\left[\int_Z^\alpha A(\zeta)d\mu(\zeta)\right]W^{-1} = \int_Y^\beta \int_Z^\alpha A(\zeta)d\mu_\xi(\zeta)d\nu(\xi)$$

and

(iii) if A is an α-Borel field of von Neumann algebras over H then $\xi \mapsto \int_Z^\alpha A(\zeta)d\mu_\xi(\zeta)$ is a β-Borel field of von Neumann algebras over $\xi \mapsto L^2(\mu_\xi;H,\alpha)$ and

$$W\left[\int_Z^\alpha A(\zeta)d\mu(\zeta)\right]W^{-1} = \int_Y^\beta \int_Z^\alpha A(\zeta)d\mu_\xi(\zeta)d\nu(\zeta)$$

Proof. Let M be the algebra of diagonalizable operators on $L^2(\mu;H,\alpha)$ and let N be the subalgebra of M consisting of the operators of the form

24. SOME TECHNICAL LEMMAS

$\int_Z^\alpha \psi(\omega(\zeta)) I_{H(\zeta)} d\mu(\zeta)$, $\psi \in L^\infty(\nu)$. Then N is a von Neumann algebra which is *-isomorphic to $L^\infty(\nu)$ and the lemma follows without any difficulty from Examples 10.9 and 20.2. □

LEMMA 24.2. Suppose that Z is a standard Borel space, that M_0 is the set of all finite nonzero Borel measures on Z, that H is a Borel field of Hilbert spaces on Z, and that α is a coherence for H. Then M_0 is a standard subset of $M(Z)$, $\mu \mapsto L^2(\mu;H,\alpha)$ is a Borel field of Hilbert spaces on M_0, and there is a coherence β for this field such that

(i) if v [respectively, A] is an α-Borel vector field [respectively, operator field] over H such that $\|v(\cdot)\|$ [respectively, $\|A(\cdot)\|$] is a bounded function on Z then $\mu \mapsto \int_Z^\alpha v(\zeta) d\mu(\zeta)$ [respectively, $\mu \mapsto \int_Z^\alpha A(\zeta) d\mu(\zeta)$] is a β-Borel vector field [respectively, operator field] over $\mu \mapsto L^2(\mu;H,\alpha)$, and

(ii) if A is an α-Borel field of von Neumann algebras over H then $\mu \mapsto \int_Z^\alpha A(\zeta) d\mu(\zeta)$ is a β-Borel field of von Neumann algebras over $\mu \mapsto L^2(\mu;H,\alpha)$.

Proof. That M_0 is a standard subset of $M(Z)$ is clear from Proposition 4.4. Let (v_j) be a sequence of α-Borel vector fields over H such that each of the functions $\|v_j(\cdot)\|$ is bounded on Z and such that the sequence $(v_j(\zeta))$ is dense in $H(\zeta)$ for each $\zeta \in Z$ and let (ϕ_k) be a sequence of Borel functions from Z to the closed unit ball in the complex plane whose image in $L^\infty(\mu)$ is weak *-dense in the unit ball of $L^\infty(\mu)$ for each $\mu \in M_0$ (cf. Example 10.9). Then the vectors $\int_Z^\alpha \phi_k(\zeta) v_j(\zeta) d\mu(\zeta)$ are total in $L^2(\mu;H,\alpha)$ for each $\mu \in M_0$ by Lemma 7.3 and, moreover, the inner product

$$\langle \int_Z^\alpha \phi_k(\zeta) v_j(\zeta) d\mu(\zeta), \int_Z^\alpha \phi_n(\zeta) v_m(\zeta) d\mu(\zeta) \rangle$$
$$= \int_Z \phi_k(\zeta) \bar{\phi}_n(\zeta) \langle v_j(\zeta), v_m(\zeta) \rangle d\mu(\zeta)$$

is a Borel function of μ on M_0 for all choices of j,k,m and n. Hence $\xi \mapsto L^2(\mu;H,\alpha)$ is a Borel field of Hilbert spaces on M_0 and there is a coherence β for this field such that each of the vector fields $\mu \mapsto \int_Z^\alpha \phi_k(\zeta) v_j(\zeta) d\mu(\zeta)$ is α-Borel (Proposition 8.1). Assertion (i) now follows readily from Lemma 8.2.

Now let A be an α-Borel field of von Neumann algebras on Z and let (A_j) be an α-Borel generating sequence for A with $\|A_j(\cdot)\| \leq 1$ for each j and each $\zeta \in Z$. Then by Theorem 18.2 the operators $\int_Z^\alpha A_j(\zeta) d\mu(\zeta)$ together with the $\int_Z^\alpha \phi_k(\zeta) I_{H(\zeta)} d\mu(\zeta)$ generate $\int_Z^\alpha A(\zeta) d\mu(\zeta)$ for each $\mu \in M_0$, and therefore $\mu \mapsto \int_Z^\alpha A(\zeta) d\mu(\zeta)$ is a β-Borel field of von Neumann algebras. □

LEMMA 24.3. Suppose that X [respectively, Y] is a standard Borel space, that μ [respectively, ν] is a Borel measure on X [respectively, Y], that H [respectively, K] is a Borel field of Hilbert spaces on X [respectively, Y], that α [respectively, β] is a coherence for H [respectively, K], and that A [respectively, B] is an α-Borel [respectively, β-Borel] field of von Neumann algebras over H [respectively, K]. Then there is a coherence γ for the field of Hilbert spaces $(\zeta,\xi) \mapsto H(\zeta) \otimes K(\xi)$ on X × Y such that the field of von Neumann algebras $(\zeta,\xi) \mapsto A(\zeta) \otimes B(\xi)$ is γ-Borel and such that the algebras

$$\int_X^\alpha A(\zeta) d\mu(\zeta) \otimes \int_Y^\beta B(\xi) d\nu(\xi) \quad \text{and} \quad \int_{X \times Y}^\gamma A(\zeta) \otimes B(\xi) d(\mu \otimes \nu)(\zeta,\xi)$$

are spatially isomorphic.

Proof. One may as well assume that the fields H and K and the coherences α and β are constant, and hence (by an abuse of notation) that H and K are fixed Hilbert spaces and that A and B are Borel maps from X to vN(H) and from Y to vN(K), respectively. It is then an easy consequence of Theorem 17.1 that $(\zeta,\xi) \mapsto A(\zeta) \otimes B(\xi)$ is a Borel map from X × Y to vN$(H \otimes K)$. It follows rather easily from Proposition 5.2 that there is a unique linear isometry W of $L^2(\mu;H) \otimes L^2(\nu;K)$ onto $L^2(\mu \otimes \nu; H \otimes K)$ with the property that

$$W[(fx)(\mu) \otimes (gy)(\nu)] = [(f \otimes g)(x \otimes y)](\mu \otimes \nu)$$

whenever $f \in L^2(\mu)$, $g \in L^2(\nu)$, $x \in H$, and $y \in K$. If A [respectively, B] is a μ-essentially bounded [respectively, ν-essentially bounded] Borel map from X to $L(H)$ [respectively, from Y to $L(K)$] satisfying $A(\zeta) \in A(\zeta)$ for each $\zeta \in X$ [respectively, $B(\xi) \in B(\xi)$ for each $\xi \in Y$] then one can use a standard approximation argument to show that

$$W \left[\int_X^\oplus A(\zeta) d\mu(\zeta) \otimes \int_Y^\oplus B(\xi) d\nu(\xi) \right] W^{-1} = \int_{X \times Y}^\oplus A(\zeta) \otimes B(\xi) d(\mu \otimes \nu)(\zeta,\xi)$$

Finally, it follows rather easily from this equality that W implements a spatial isomorphism between the two von Neumann algebras in question. □

24. SOME TECHNICAL LEMMAS

LEMMA 24.4. Suppose that H is a separable infinite-dimensional Hilbert space and that S is a Borel subset of vN(H) satisfying $a(S) = S$. Let $\hat{S}$ denote the set of all those von Neumann algebras on H which are *-isomorphic to $\int_Z^\oplus A(\zeta)d\mu(\zeta)$ for some standard Borel space Z, some nonzero Borel measure μ on Z, and some Borel map A from Z to vN(H). Then $\hat{S}$ is an analytic subset of vN(H).

Proof. Let M_0 be the set of all finite nonzero Borel measures on $Y = S \times \mathbb{N}$ and let $M_1, M_2, \ldots$ be as in the proof of Corollary 22.5. Then $L^2(\nu; H \otimes H)$ is separable and infinite-dimensional for each $\nu \in M_0$ by Proposition 5.2 and Corollary 5.3, and hence there is (by Lemma 24.2) a Borel map B from M_0 to vN(H) such that

$$B(\nu) \simeq \int_Y^\oplus A \otimes M_n d\nu(A,n), \quad \nu \in M_0$$

here and throughout this proof, "$\simeq$" means "is *-isomorphic to." Now M_0 is a standard subset of the Borel space $M(Y)$ (see Lemma 24.2) and hence $B(M_0)$, the range of B, is an analytic subset of vN(H). But then $a(B(M_0))$ too is an analytic subset of vN(H) by Lemma 22.4, and so to prove the lemma it is sufficient to show that $a(B(M_0)) = \hat{S}$.

For each measure ν in M_0 and each positive integer n, let ν_n be the Borel measure $S \mapsto \nu(S \times \{n\})$ on S and let $J(\nu) = \{n \in \mathbb{N}: \nu_n \neq 0\}$. Then clearly

$$B(\nu) \simeq \oplus_{n \in J(\nu)} \int_S^\oplus A \otimes M_n d\nu_n(A)$$

and hence

$$B(\nu) \simeq \oplus_{n \in J(\nu)} M_n \otimes \int_S^\oplus A d\nu_n(A) \tag{1}$$

by Proposition 19.7. For each $n \in J(\nu)$ there is (by Theorem 2.14, Example 20.1, and Lemma 24.3) a finite Borel measure μ_n on the closed unit interval X, a Borel field of Hilbert spaces H_n on X, and a coherence α_n for the field of Hilbert spaces $(\zeta, A) \mapsto H_n(\zeta) \otimes H$ on $X \times S$ such that the algebras $M_n \otimes \int_S^\oplus A d\nu_n(A)$ and

$$\int_{X \times S}^{\alpha_n} C(H_n(\zeta)) \otimes A d(\mu_n \otimes \nu_n)(\zeta, A)$$

are spatially isomorphic. Now let θ be some linear isometry of ℓ^2 onto H, let $Z = X \times S \times J(\nu)$, and define a Borel measure λ on Z by putting

$$\lambda(S) = \sum_{n \in J(\nu)} (\mu_n \otimes \nu_n)(\{(\zeta,A) \in X \times S: (\zeta,A,n) \in S\})$$

for each Borel subset S of Z. For each point $(\zeta,A,n) \in Z$ it is the case that $\beta(\zeta,A,n) = \alpha_n(\zeta,A)$ is onto ℓ^2 and hence that $\theta \circ \alpha_n(\zeta,A)$ is a linear isometry of $H_n(\zeta) \otimes H$ onto H. Thus

$$(\alpha,A,n) \mapsto [\theta \circ \alpha_n(\zeta,A)][C(H_n(\zeta)) \otimes A][\theta \circ \alpha_n(\zeta,A)]^{-1}$$

is a Borel map from Z to S and

$$B(\nu) \simeq \oplus_{n \in J(\nu)} \int_{X \times S}^{\alpha_n} C(H_n(\zeta)) \otimes Ad(\mu_n \otimes \nu_n)(\zeta,A)$$

$$\simeq \int_Z^\beta C(H_n(\zeta)) \otimes Ad\lambda(\zeta,A,n)$$

$$\simeq \int_Z^\theta [\theta \circ \alpha_n(\zeta,A)][C(H_n(\zeta)) \otimes A][\theta \circ \alpha_n(\zeta,A)]^{-1} d\lambda(\zeta,A,n)$$

by Theorem 19.8. This shows that each $B(\nu)$ belongs to $\hat{S}$, and hence that $a(B(M_0)) \subset \hat{S}$.

Turning to the proof of the inclusion $\hat{S} \subset a(B(M_0))$, one must show that corresponding to any standard Borel space Z, any nonzero Borel measure μ on Z, and any Borel map A from Z to S there is a measure ν in M_0 with

$$\int_Z^\theta A(\zeta) d\mu(\zeta) \simeq B(\nu) \tag{2}$$

Accordingly, say that one is given such Z, μ and A, and assume (as one may by Lemma 7.2) that μ is finite. Let $\sim$ be the equivalence relation on Z under which two points are equivalent if and only if their images under A are equal, give $T = Z/\sim$ the quotient Borel structure, and let ω be the canonical mapping of Z onto T. Then there is a unique map $\tilde{A}$ from T to S satisfying $\tilde{A} \circ \omega = A$, and this map is one-one and Borel. But then T will be countably separated as S is, and therefore analytic by Theorem 3.1. Put $\tilde{\mu} = \omega_*(\mu)$. From Corollary 4.2 and Theorem 4.3 and the fact that one can delete a given $\tilde{\mu}$-null Borel set from T by deleting a suitable μ-null Borel set from Z, one may assume that T is standard and that there is a Borel mapping σ from T to Z satisfying $\omega(\sigma(\xi)) = \xi$ for all $\xi \in T$. Let $\xi \mapsto \mu_\xi$ be the Borel map from T to $M(Z)$ obtained by applying Theorem 4.5 to Z, T, ω, and μ. Then by Lemma 24.1 there is a coherence α for the field of Hilbert spaces $\xi \mapsto L^2(\mu_\xi; H)$ on T such that the field of von Neumann algebras $\xi \mapsto \int_Z^\theta A(\zeta) d\mu_\xi(\xi)$ is α-Borel and such that

24. SOME TECHNICAL LEMMAS

$$\int_Z^\oplus A(\zeta)d\mu(\zeta) \simeq \int_T^\alpha \int_Z^\oplus A(\zeta)d\mu_\xi(\zeta)d\tilde{\mu}(\xi)$$

Moreover, Lemmas 22.4 and 24.1, the latter this time being applied to the constant field $\zeta \mapsto C$, together imply that the sets

$$T_n = \{\xi \in T: L^\infty(\mu_\xi) \simeq M_n\}, \quad n \in \mathbb{N}$$

are all Borel. For each positive integer n let $\tilde{\mu}_n$ denote the Borel measure $S \mapsto \tilde{\mu}(S \cap T_n)$ on T and let $J = \{n \in \mathbb{N}: \tilde{\mu}_n \neq 0\}$. Then for each $n \in \mathbb{N}$ and each $\xi \in T_n$ one has

$$\int_Z^\oplus A(\zeta)d\mu_\xi(\zeta) \simeq A(\sigma(\xi)) \otimes M_n = \tilde{A}(\xi) \otimes M_n$$

by Corollary 19.9, and therefore

$$\int_Z^\oplus A(\zeta)d\mu(\zeta) \simeq \oplus_{n \in J} \int_T^\oplus A(\xi) \otimes M_n d\tilde{\mu}_n(\xi) \simeq \oplus_{n \in J} M_n \otimes \int_T^\oplus \tilde{A}(\xi)d\tilde{\mu}_n(\xi) \quad (3)$$

by Proposition 19.7 and Theorem 19.8.

For each positive integer n let p_n denote the map $\xi \mapsto (\tilde{A}(\xi),n)$ from T to $Y = S \times \mathbb{N}$. Then $\nu = \sum_{n=1}^\infty p_{n*}(\tilde{\mu}_n)$ belongs to M_0 and is such that $\nu_n = \tilde{A}_*(\tilde{\mu}_n)$ for each n in $\mathbb{N}$ and such that $J(\nu) = J$. Now Corollary 2.12 and Theorem 2.13 imply that $\tilde{A}(T)$ is a Borel subset of S and that $\tilde{A}$ is a Borel isomorphism of T onto $\tilde{A}(T)$, and therefore

$$\int_S^\oplus Ad\nu_n(A) \simeq \int_T^\oplus \tilde{A}(\xi)d\tilde{\mu}_n(\xi), \quad n \in J \tag{4}$$

The desired isomorphism (2) now results from simply combining the isomorphisms (1), (3), and (4). □

This completes the proof of the first of the two technical lemmas on which the proof of Theorem 21.1 (i) will depend. Turning now to the other one, it will be convenient to begin by fixing some notation. Consider the horizontal strip S consisting of those complex numbers z satisfying $0 \leq \text{Im}(z) \leq 1$, and let F [respectively, $C_0(S)$] denote the collection of all those bounded continuous complex-valued functions on S which are analytic in the interior of S [respectively, which vanish at infinity]. The sup-norm topology on $C_0(S)$ is separable and is induced by a complete metric, and hence generates a standard Borel structure. Consider the map from F to

$C_0(S)$ which sends an f in F onto the function $z \mapsto f(z)\exp(-|\text{Re}(z)|)$: this map is injective, carries each of the sets

$$\{f \in F: |f(z)| \leq a \text{ for all } z \in S\}, \quad a > 0$$

onto a sup-norm closed subset of $C_0(S)$, and hence carries F onto a Borel subset of $C_0(S)$. The weakest Borel structure on F making the map in question Borel is therefore a standard one with the property that the evaluation maps $f \mapsto f(z)$, $z \in S$, are all Borel.

It will be necessary to know that the maximum modulus principle is valid for functions in F, i.e., that if ∂S denotes the boundary of S then $\sup_{z \in S} |f(z)| = \sup_{z \in \partial S} |f(z)|$ for all $f \in F$. To see this, let f be some function in F and put $M = \sup_{z \in S} |f(z)|$ and $N = \sup_{z \in \partial S} |f(z)|$. Let ε be some given positive number and for each complex number z not on the nonpositive real axis let $z^{-\varepsilon}$ denote the principal value of the $(-\varepsilon)^{\text{th}}$ power of z. Then one has

$$|(iz + 2)^{-\varepsilon}| \leq (4 + [\text{Re}(z)]^2)^{-\varepsilon/2} \leq 1$$

for all $z \in S$. Now let a be some positive number with $(4 + a^2)^{-\varepsilon/2} \leq NM^{-1}$ and let $S_a = \{z \in S: |\text{Re}(z)| \leq a\}$. Then the function $z \mapsto (iz + 2)^{-\varepsilon} f(z)$ is continuous on S_a, is analytic on the interior of S_a, and is bounded by N on the boundary of S_a; thus $|(iz + 2)^{-\varepsilon} f(z)| \leq N$ for all $z \in S_a$ by the usual maximum modulus principle. Now this conclusion is actually valid for all sufficiently large numbers a, and therefore $|(iz + 2)^{-\varepsilon} f(z)| \leq N$ for all $z \in S$. Noticing that this conclusion, in turn, is valid for all positive numbers ε and that $\lim_{\varepsilon \downarrow 0} |(iz + 2)^{-\varepsilon}| = 1$ for each fixed point z in S, it must be the case that $|f(z)| \leq N$ for all $z \in S$.

Let H be a separable infinite-dimensional Hilbert space, let $U(H)$ be the group of unitary operators on H, and let $U_0(H)$ be the subset of $U(H)$ consisting of those operators for which 1 is not an eigenvalue. Recall from Lemma 22.3 that $U(H)$ is a Polish group; give $U_0(H)$ the relative topology and Borel structure. Also recall that the Cayley transform is the injective map $A \mapsto (A + iI)(A - iI)^{-1}$ from the set of all (not necessarily bounded) self-adjoint operators on H onto $U_0(H)$. Let T be the unit circle in the complex plane and let $T_0 = T - \{1\}$. For each real number t let g_t denote the function $z \mapsto \exp[-t(z + 1)(z - 1)^{-1}]$ from T_0 to T. Then for each fixed W in $U_0(H)$, $g_t(W)$ exists and is a unitary operator on H for each real t;

24. SOME TECHNICAL LEMMAS

$t \mapsto g_t(W)$ is a strongly continuous unitary group; and W is the Cayley transform of the infinitesimal generator of this unitary group. Indeed, the strong continuity follows from spectral theory and Proposition C.3, and the third assertion follows from the fact that $A = i(W + I)(W - I)^{-1}$ is a self-adjoint operator on H whose Cayley transform is W and which satisfies $\exp(itA) = g_t(W)$ for all real t. For each fixed t, on the other hand, g_t is a pointwise limit of a uniformly bounded sequence of polynomials in z and $\bar{z}$, and hence $W \mapsto g_t(W)$ is a Borel map from $U_0(H)$ to $U(H)$ (this too depends on spectral theory). In view of Proposition 2.3 this means, in particular, that the map $(t,W) \mapsto g_t(W)$ from $\mathbb{R} \times U_0(H)$ to $U(H)$ is Borel.

LEMMA 24.5. $U_0(H)$ is a G_δ-subset of $U(H)$.

Proof. Fix a dense sequence of vectors (x_j) in H and a sequence of continuous functions (f_m) from T to the closed unit interval satisfying $f_m(z) = 1$ if $z \in T$ and $|z - 1| \leq 2^{-m}$ and $f_m(z) = 0$ if $z \in T$ and $|z - 1| \geq 2^{-m+1}$. Consider a unitary operator U on H, let $E(\cdot)$ be its spectral measure (which is defined on the Borel subsets of T), and assume that 1 is not an eigenvalue of U. Then $E(\{1\}) = 0$. Put

$$P_m = E(\{z \in T: |z - 1| \leq 2^{-m}\}), \quad m \in \mathbb{N}$$

and notice that $P_1 \geq P_2 \geq \ldots$, that $0 \leq f_{m+1}(U) \leq P_m$ for $m \in \mathbb{N}$, and that $P_m \to 0$ strongly as $m \to \infty$. Hence given any positive integer n, one can find an integer m with $\|f_m(U)x_j\| < \frac{1}{2}$ for $1 \leq j \leq n$. Now suppose that, conversely, 1 is an eigenvalue of U. Then there is a unit vector x in H with $Ux = x$, and hence with $f_m(U)x = x$ for all m. Letting n be an integer with $\|x - x_n\| < \frac{1}{4}$, one has

$$\|f_m(U)x_n\| \geq 1 - |1 - \|f_m(U)x_n\||$$

$$= 1 - |\|f_m(U)x\| - \|f_m(U)x_n\||$$

$$\geq 1 - \|f_m(U)(x - x_n)\|$$

$$\geq \tfrac{3}{4}$$

for all m.

It should now be clear that one has the identity

$$U_0(H) = \bigcap_{n=1}^{\infty} \bigcup_{m=1}^{\infty} \bigcap_{j=1}^{n} \{U \in U(H): \|f_m(U)x_j\| < \tfrac{1}{2}\}$$

So to prove the lemma it will be sufficient to show that the map $U \mapsto \|f(U)x\|$ on $U(H)$ is continuous for any continuous real-valued function f on T and any vector x in H. But this is clear as such an f is a uniform limit of polynomials in z and $\bar{z}$ and as $U \mapsto U^*$ is strongly continuous on $U(H)$ by Lemma 22.3. □

LEMMA 24.6. The two sets

$$S_1 = \{(A,W) \in vN_g(H) \times U(H): A = WAW^*\}$$

and

$$S_2 = \{(A,W) \in S_1: A \mapsto WAW^* \text{ is an outer automorphism of } A\}$$

are both Borel subsets of $vN_g(H) \times U(H)$.

Proof. Choose a metric d on the unit ball of $L(H)$ which induces the weak operator topology as well as a sequence (A_j) of Borel functions from $vN(H)$ to $L(H)$ such that for each A in $vN(H)$ the sequence $(A_j(A))$ is weakly dense in the unit ball of A (such a sequence exists by Theorem 17.1). Then a point (A,W) in $vN_g(H) \times U(H)$ lies in S_1 if and only if

$$WA_j(A)W^*A_k(A') = A_k(A')WA_j(A)W^*$$

for all j and k, and so S_1 is Borel by Theorem 17.1 (e). For any four positive integers j, k, m and n let $S(j,k,m,n)$ denote the subset of S_1 consisting of those pairs (A,W) for which

$$d(A_j(A),0) \geq n^{-1} \quad \text{and} \quad d(A_j(A),WA_k(A')) \leq m^{-1}$$

Then Theorem 17.1 (e) implies that

$$S = \cup_n \cap_m \cup_{j,k} S(j,k,m,n)$$

is a Borel subset of S_1, and so it will be sufficient to show that $S = S_1 - S_2$.

Consider first a point (A,W) in $S_1 - S_2$. There is, by assumption, a unitary operator U in A with $UAU^* = WAW^*$ for all $A \in A$, or equivalently, with $U \in WA'$. Choose an n in $\mathbb{N}$ with $d(U,0) \geq 2n^{-1}$, and let m be an arbitrary positive integer. Then one can find positive integers j and k such that

24. SOME TECHNICAL LEMMAS

$$d(U, A_j(A)) \leq (2mn)^{-1} \quad \text{and} \quad d(U, WA_k(A')) \leq (2mn)^{-1}$$

It now follows from the triangle inequality that (A,W) belongs to $S(j,k,m,n)$; and since such j and k can be found corresponding to any given m in $\mathbb{N}$, it must actually be the case that (A,W) belongs to S.

Now suppose that, conversely, (A,W) is a point in S. Then (A,W) belongs to S_1 by the definition of S, and so one must show that $A \cap WA'$ contains a unitary operator. Now by assumption there is a positive integer n and two functions j and k from $\mathbb{N}$ to itself such that

$$(A,W) \in S(j(m), k(m), m, n), \quad m \in \mathbb{N}$$

Letting B denote some weak limit point of the sequence $(A_{j(m)}(A))$ (such a limit point must exist by the weak compactness of the unit ball of A), one must have $B \in A$ and $0 \neq B \in WA'$. But then $WAW^*B = BA$ for all $A \in A$, and therefore $B^*B = A^*B^*BA$ and

$$BB^* = (WAW^*)BB^*(WAW^*)^*$$

for all $A \in A \cap U(H)$. This implies that both B^*B and BB^* belongs to A', and hence that $B^*B = BB^* = \|B\|^2 I$ as A is a factor. From the polar decomposition of B one now deduces that $B = \|B\|U$ for some unitary operator U which must lie in A. □

LEMMA 24.7. *The set* $vN_3(H) \cap vN_9(H)$ *is Borel.*

Proof. As $vN_9(H)$ is Borel (by Theorem 17.1) and as $vN_3(H)$ is coanalytic (by Corollary 22.13), the set in question must be a coanalytic subset of $vN(H)$. So to prove the lemma it is sufficient (by Corollary 2.9) to show that $vN_3(H) \cap vN_9(H)$ is an analytic subset of $vN(H)$.

Fix once and for all a unit vector x_0 in H, and let ϕ be the vector state $A \mapsto \langle Ax_0, x_0 \rangle$ on $L(H)$ defined by x_0 and S the set of all those factors of type III on H for which x_0 is a cyclic and separating vector. Suppose that A is a factor of type III on H and let E be the projection onto the smallest closed subspace of H containing $A'x_0$. Then E belongs to A and is equivalent to I in A, i.e., there is a partial isometry V in A with $V^*V = I$ and $VV^* = E$. Put $x_1 = V^*x_0$. Then x_1 is a unit vector and $A'x_1 = V^*A'x_0$ is dense in H. Now let F be the projection onto the smallest closed subspace of H containing Ax_1. Then F belongs to A' and is equivalent to I in

A' by Theorem A.9, i.e., there is a partial isometry W in A' with $W^*W = I$ and $WW^* = F$. Put $x_2 = W^*x_1$. Then x_2 is a unit vector and $Ax_2 = W^*Ax_1$ is dense in H, and, moreover, $A'x_2$ is dense in H as $A'x_2 \supset A'WW^*x_1 = A'x_1$. The vector x_2 is therefore both cyclic and separating for A. So if U is any unitary operator on H with $Ux_2 = x_0$ then x_0 will be a cyclic and separating vector for UAU^*, i.e., $UAU^* \in S$. This shows that $s(S) = vN_3(H) \cap vN_9(H)$, and so to prove the lemma it will be enough (by Lemma 22.4) to show that S is an analytic subset of $vN(H)$.

Let (A_j) be a sequence of Borel maps from $vN(H)$ to $L(H)$ with the property that for each A in $vN(H)$ the sequence $(A_j(A))$ is strongly dense in A. For any two positive integers j and k let $S(j,k)$ be the subset of $vN_9(H) \times U_0(H) \times F \times \mathbb{R}$ consisting of those points (A,W,f,s) for which

(i) x_0 is a cyclic and separating vector for A,
(ii) $g_t(W)x_0 = x_0$ for all real t,
(iii) $g_t(W)Ag_{-t}(W) = A$ for all real t,
(iv) the automorphism $A \mapsto g_s(W)Ag_{-s}(W)$ of A is outer,
(v) $f(t) = \phi(g_t(W)A_j(A)g_{-t}(W)A_k(A))$ for all real t, and
(vi) $f(t + i) = \phi(A_k(A)g_t(W)A_j(A)g_{-t}(W))$ for all real t.

If E_0 denotes the projection onto the one-dimensional subspace of H spanned by x_0 then, in the notation of Section 23, x_0 is a cyclic and separating vector for an A in $vN(H)$ if and only if $P(A,E_0) = P(A',E_0) = I$. So by Lemma 23.1 and Theorem 17.1 the subset of $vN(H)$ consisting of those A satisfying (i) is Borel. Next, notice that by continuity the phrase "for all real t" in (ii), (iii), (v), and (vi) could be replaced by "for all rational t" without altering the effect of these conditions. This remark, the fact that $(t,W) \mapsto g_t(W)$ is a Borel map from $\mathbb{R} \times U_0(H)$ to $U(H)$, and Lemma 24.6 together show that the set consisting of those points (A,W,f,s) in $vN_9(H) \times U_0(H) \times F \times \mathbb{R}$ satisfying (ii)-(vi) is Borel. This shows that each of the sets $S(j,k)$ is a Borel subset of $vN_9(H) \times U_0(H) \times F \times \mathbb{R}$, and hence is a standard Borel space in its relative Borel structure by Lemma 24.5. Now let p and q denote the projections of $vN_9(H) \times U_0(H) \times F \times \mathbb{R}$ onto $vN_9(H) \times U_0(H) \times \mathbb{R}$ and of $vN_9(H) \times U_0(H) \times \mathbb{R}$ onto $vN_9(H)$, respectively, and put

$$S_0 = q(\cap_{j,k=1}^{\infty} p(S(j,k)))$$

Then S_0 is an analytic subset of $vN(H)$, and so the lemma would be proven if one could show that $S = S_0$.

24. SOME TECHNICAL LEMMAS

Suppose that A is an element of S, and let (σ_t) be the modular automorphism group of A associated with the restriction of ϕ to A (cf. Appendix A). Then σ_s must be outer for some $s \in \mathbb{R}$. Since x_0 is a cyclic vector for A and since

$$\|\sigma_t(A)x_0\|^2 = \phi(\sigma_t(A^*A)) = \phi(A^*A) = \|Ax_0\|^2$$

for all $A \in A$ and all $t \in \mathbb{R}$, one can construct a strongly continuous one-parameter unitary group (U_t) on H with $U_t A x_0 = \sigma_t(A) x_0$ and $U_t A U_{-t} = \sigma_t(A)$ for all $A \in A$ and all $t \in \mathbb{R}$. Letting W be the Cayley transform of the infinitesimal generator of this unitary group, one knows that $W \in U_0(H)$ and that $g_t(W) = U_t$ for all real t. Remembering that (σ_t) satisfies the KMS-condition, it is clear that (A,W,s) lies in $\cap_{j,k} p(S(j,k))$, and hence that A belongs to S_0.

Now suppose that, conversely, A is an element of S_0. Then A is certainly a factor which has x_0 as a cyclic and separating vector. The definition of S_0 implies that there is a $W \in U_0(H)$ and an $s \in \mathbb{R}$ such that

$$(A,W,s) \in \cap_{j,k=1}^{\infty} p(S(j,k))$$

Putting

$$\sigma_t(A) = g_t(W) A g_{-t}(W), \quad A \in A \quad \text{and} \quad t \in \mathbb{R}$$

one obtains a strongly continuous one-parameter automorphism group of A. Given two operators B and C in A, there will be maps a and b from $\mathbb{N}$ to itself such that $A_{a(n)} \to B$ and $A_{b(n)} \to C$ strongly as $n \to \infty$. But then as (A,W,s) belongs to $S(a(n),b(n))$ for each n there will, in turn, be a sequence (f_n) in F for which the formulae

$$f_n(t) = \phi(\sigma_t(A_{a(n)}) A_{b(n)})$$

and

$$f_n(t + i) = \phi(A_{b(n)} \sigma_t(A_{a(n)}))$$

hold for all $t \in \mathbb{R}$ and $n \in \mathbb{N}$. But then

$$|f_m(t) - f_n(t)| \leq |\phi(\sigma_t(A_{a(m)})(A_{b(m)} - A_{b(n)}))|$$
$$+ |\phi(\sigma_t(A_{a(m)} - A_{a(n)})A_{b(n)}))|$$
$$= |<(A_{b(m)} - A_{b(n)})x_0, \sigma_t(A_{a(m)}^*)x_0>|$$
$$+ |<A_{b(n)}x_0, \sigma_t((A_{a(m)} - A_{a(n)})^*)x_0>|$$
$$\leq \|(A_{b(m)} - A_{b(n)})x_0\| \|A_{a(m)}x_0\|$$
$$+ \|A_{b(n)}x_0\| \|A_{a(m)} - A_{a(n)})x_0\|$$

for all $t \in \mathbb{R}$ and $m,n \in \mathbb{N}$, and therefore $|f_m(t) - f_n(t)| \to 0$ uniformly in t as $m,n \to \infty$. A similar calculation will show that $|f_m(t+i) - f_n(t+i)| \to 0$ uniformly in t as $m,n \to \infty$. The maximum modulus for functions in F now implies that the sequence (f_n) converges uniformly on compact subsets of S to a function, say f, in F. This limit f clearly satisfies the formulae

$$f(t) = \phi(\sigma_t(B)C) \quad \text{and} \quad f(t+i) = \phi(C\sigma_t(B))$$

for all real t. The uniqueness of modular automorphism groups now implies that (σ_t) is in fact the modular automorphism group of A associated with the restriction of ϕ to A, and then the fact that σ_s is outer implies that A is of type III. This shows that A belongs to S and completes the proof of the equality $S = S_0$. □

25. The Completion of the Proof of Theorem 21.1

With the results of the last section now available one can complete the proof of Theorem 21.1 easily and quickly. But before actually doing so it might be useful to recall at what stage the proof is presently at. Letting H be some separable infinite-dimensional Hilbert space, it was shown in Section 22 that part (ii) of Theorem 21.1 follows from part (i), that the sets $vN_{1,n}(H)$, $n \in \mathbb{N} \cup \{\infty\}$, are all Borel, and that $vN_k(H)$ is a Borel [respectively, an analytic, a coanalytic] subset of $vN(H)$ for $k = 1,5,8,9$ [respectively, $k = 6$, $k = 3,4,7$]. In completing the proof of Theorem 21.1 it will be necessary to make use of Theorem 21.2 (i), and one may do this as the latter theorem was proven in Section 23.

Notice that each of the sets $vN_k(H) \cap vN_9(H)$, $1 \leq k \leq 7$, is Borel by Corollary 22.5 and Lemmas 22.8 and 24.7. Suppose for the moment one knew that

25. THE COMPLETION OF THE PROOF OF THEOREM 21.1

$$[vN_k(H) \cap vN_9(H)]^{\wedge} = vN_k(H), \quad 2 \leq k \leq 7 \tag{*}$$

where $\wedge$ is as in Lemma 24.4. Then it would follow from Lemma 24.4 that each of the sets $vN_k(H)$, $2 \leq k \leq 7$, is analytic and, in turn, from Corollaries 2.9, 22.6, 22.11 and 22.13 that $vN_k(H)$ is a Borel set for $k = 3, 4, 7$. If ω is as in the proof of Corollary 22.6 then $a(\omega(vN(H) \times vN_3(H)))$ is just the complement of $vN_6(H)$ in $vN(H)$, and therefore $vN_6(H)$ would be a Borel set by Lemma 22.4 and Corollary 22.9. Finally, it is now clear that $vN_2(H) = vN_4(H) \cap vN_6(H)$ too would be a Borel set. So to complete the proof of Theorem 21.1 it only remains to verify (*).

To prove (*) it is in fact only necessary to judiciously combine a number of preceeding results. First of all, if μ is a Borel measure on a standard Borel space Z and if A is a Borel map from Z into $vN_k(H) \cap vN_9(H)$ then $\int_Z^{\oplus} A(\zeta) d\mu(\zeta)$ is P_k by Theorem 22.2 (i). On the other hand, if A_0 is an element of $vN_k(H)$ then by Theorems 19.4 and 19.6 and Proposition 19.7 there will be a standard Borel space Z, a Borel measure μ on Z, a Borel field of infinite-dimensional Hilbert spaces K on Z, a coherence α for K, and an α-Borel field of factors A over K such that $A_0 \otimes C(\ell^2)$ and $\int_Z^{\alpha} A(\zeta) d\mu(\zeta)$ are spatially isomorphic. Moreover, one may as well assume that both K and α are constant, that H is the value of the constant function K, and finally (by Theorem 22.2 (i)) that $A(\zeta) \in vN_k(H) \cap vN_9(H)$ for each $\zeta \in Z$. But this just means that A_0 belongs to $[vN_k(H) \cap vN_9(H)]^{\wedge}$.

HISTORICAL COMMENTS

One can formulate Theorem 21.1 so as not to involve the Effros Borel structure: the inverse image of the sets $vN_k(H)$ and $vN_{1,n}(H)$, $1 \leq k \leq 9$ and $1 \leq n \leq \infty$, under the composition of the functions E_k and $E_{1,n}$, $1 \leq k \leq 7$ and $1 \leq n \leq \infty$, with a map from a standard Borel space into $vN(H)$ satisfying condition (d) of Theorem 17.1 must be Borel. Theorem 21.1 in this form and Theorem 21.2 go back to von Neumann's paper [49, Section 22] and to an early paper of Dixmier [5, Section IV], respectively, and are contained in Dixmier's book [10]. In fact, von Neumann proved that $vN_9(H) \cap W$ is a universally measurable subset of $vN_9(H)$ if W is any one of the sets $vN_{1,n}(H)$, $1 \leq n \leq \infty$, $vN_1(H)$, $vN_5(H)$, $vN_2(H) \cap vN_5(H)$, and $vN_2(H) \cap vN_3(H)$, and Dixmier proved certain parts of Theorem 21.2 (i) in the case where the decomposition in question is central. Both of these authors encountered measure-theoretic difficulties in distinguishing between factors of types II_{∞} and III (cf. the first paragraph of Section 24). These difficulties were overcome by Schwartz who showed that $vN_9(H) \cap vN_2(H)$ is a universally measurable subset of $vN_g(H)$ and completed the proof of the special case of Theorem 21.2 (i) in which the decomposition is central [53].

The results contained in this chapter are for the most part due to Effros [16] and Nielsen [51]. In particular, Lemmas 22.4, 24.1-24.3, Corollary 22.5, and Proposition 22.8 can be found in [16] while the construction described towards the end of Section 22, Lemmas 22.12, 24.4, 24.6, and 24.7, and Corollary 22.14 appear in [51]. Lemma 22.3 is due to Dixmier [8, Lemme 4] and Lemma 24.5 to Kallman [30, Proposition 2.7]. The proof of Lemma 24.6 contains within it a criterion for an automorphism of a factor to be inner, a criterion which is due independently to Bures [4, Lemma 5.4] and Kallman [29, Corollary 1.2]. The fact that the functions in F satisfy the maximum modulus principle is actually a consequence of the Phragmen-Lindelöf theorem; the elementary proof given here is taken from Dunford and Schwartz's book [11, p. 231].

It is possible to give a proof of the crucial Lemma 24.7 which avoids the use of modular automorphism groups (cf. the Historical Comments following the next chapter).

Chapter 6
MEASURES AND REPRESENTATIONS

26. The Dual and the Quasi-dual

It turns out that there are several intimate relationships between representations of a separable involutive Banach algebra and measures on its dual and quasi-dual. The precise form that these relations take is that there are one-one correspondences between certain classes of representations and certain classes of measures. A hint that such correspondences might exist is already provided by Theorem 13.3.

Let R be a separable involutive Banach algebra. For $n \in \mathbb{N} \cup \{\infty\}$ one can consider the set $\text{Rep}_n R$ of all representations of R on ℓ_n^2 and its two subsets $\text{Irr}_n R$ and $\text{Fac}_n R$ consisting of the irreducible and the factor representations, respectively, in $\text{Rep}_n R$. In addition, one can consider the two sets $\text{Irr } R = \cup_{1 \leq n \leq \infty} \text{Irr}_n R$ and $\text{Fac } R = \cup_{1 \leq n \leq \infty} \text{Fac}_n R$ as well as the set of equivalence classes $\hat{R} = \text{Irr } R/\simeq$ and the set of quasi-equivalence classes $\widetilde{R} = \text{Fac } R/\approx$; $\hat{R}$ is called the *dual* of R and $\widetilde{R}$ the *quasi-dual* of R. Again for $n \in \mathbb{N} \cup \{\infty\}$, one can give $\text{Rep}_n R$ the weakest topology making the maps $\pi \mapsto \langle \pi(R)x,y \rangle$, $R \in R$ and $x,y \in \ell_n^2$, continuous as well as the Borel structure generated by this topology, $\text{Irr}_n R$ and $\text{Fac}_n R$ the relative Borel structures, $\text{Irr } R$ and $\text{Fac } R$ the sum Borel structures, and $\hat{R}$ and $\widetilde{R}$ the quotient Borel structures. These Borel structures on $\hat{R}$ and $\widetilde{R}$ are the so-called *Mackey Borel structures*. One says that R *has a smooth dual* if $\hat{R}$ is analytic; by a deep theorem of Glimm, this is equivalent to saying that R if *of type* I, i.e., that every representation of R is of type I.

THEOREM 26.1. For each $n \in \mathbb{N} \cup \{\infty\}$ $\text{Rep}_n R$ is a Polish space, the map $\pi \mapsto \pi(R)"$ from $\text{Rep}_n R$ into $vN(\ell_n^2)$ is Borel, and the sets $\{\pi \in \text{Rep}_n R : \pi(R)" \in vN_k(\ell_n^2)\}$, $1 \leq k \leq 9$, are all Borel.

Proof. The second and third assertions are obviously consequences of Theorems 17.1 and 21.1. To prove the first assertion, let n be a fixed element of $\{\infty\} \cup \mathbb{N}$, let S be a countable dense subset of R which is a *-algebra over $\mathbb{Q} + i\mathbb{Q}$, and for each non-negative real number r put

$$L(\ell_n^2)_r = \{A \in L(\ell_n^2) : \|A\| \leq r\}$$

Then $L(\ell_n^2)_r$ is a Polish space in the weak operator topology for each $r \in [0, \infty)$ and $W = \Pi_{R \in S} L(\ell_n^2)_{\|R\|}$ is Polish in the product topology. The function which associates to each element of $\text{Rep}_n R$ its restriction to S is (by Proposition B.1) a homeomorphism of $\text{Rep}_n R$ onto $W_1 \cap W_2$, where

$$W_1 = \{\rho \in W : \rho \text{ is } (\mathbb{Q} + i\mathbb{Q})\text{-linear and *-preserving}\}$$

and

$$W_2 = \{\rho \in W : \rho \text{ is multiplicative}\}$$

and as W_1 is clearly a closed subset of W it will (by Proposition 2.1) suffice to show that W_2 is a G_δ-subset of W. For each point (R,S) in $S \times S$ let $\theta_{R,S}$ denote the map

$$(\pi, \rho) \mapsto \pi(RS) - \pi(R)\rho(S)$$

from $W \times W$ to $L(\ell_n^2)$. This map is certainly separately continuous, so if Δ denotes the diagonal in $W \times W$ and p the projection of $W \times W$ onto its first coordinate then $p(\Delta \cap \theta_{R,S}^{-1}(0))$ is a G_δ-subset of W by Proposition 2.3. Indeed, if (W_j) is a sequence of open subsets of $W \times W$ with $\theta_{R,S}^{-1}(0) = \cap_j W_j$ then

$$p(\Delta \cap \theta_{R,S}^{-1}(0)) = \cap_j p(\Delta \cap W_j)$$

and if (U_k) and (V_k) are two families of open subsets of W then

$$p(\Delta \cap [\cup_k (U_k \times V_k)]) = \cup_k (U_k \cap V_k)$$

But W_2 is just the intersection of the sets $p(\Delta \cap \theta_{R,S}^{-1}(0))$ over all pairs (R,S) in $S \times S$, and hence W_2 is a G_δ-subset of W. □

26. THE DUAL AND THE QUASI-DUAL

COROLLARY 26.2. The Borel spaces Irr R and Fac R are standard.

Proof. This corollary is clear from the Theorem 21.1. □

Notice that the following proposition is analogous to Lemma 22.4.

PROPOSITION 26.3. One-point subsets of both $\hat{R}$ and $\check{R}$ are Borel.

Proof. For each $n \in \mathbb{N} \cup \{\infty\}$ let $U(\ell_n^2)$ be the group of unitary operators on ℓ_n^2 and consider the map $(U,\pi) \mapsto U\pi(\cdot)U^*$ from $U(\ell_n^2) \times \text{Rep}_n R$ to $\text{Rep}_n R$. By an argument analogous to one in the proof of Lemma 22.4 one can show that for each representation π in $\text{Rep}_n R$ the set $\{\rho \in \text{Rep}_n R: \rho \simeq \pi\}$ is Borel. Next, for each such n let W_n be some linear isometry of $\ell_n^2 \otimes \ell^2$ onto ℓ^2 and consider the map θ from Fac R to $\text{Fac}_\infty R$ defined by $\theta(\pi) = W_n[\pi(\cdot) \otimes I]W_n^*$, $\pi \in \text{Fac}_n R$. This map θ is Borel and is such that two representation π and ρ in Fac R are quasi-equivalent if and only if the representations $\theta(\pi)$ and $\theta(\rho)$ are equivalent (Proposition A.3). The proposition now follows from the facts that if p and q denote the canonical maps of Irr R onto $\hat{R}$ and of Fac R onto $\check{R}$, respectively, then

$$p^{-1}(p(\pi)) = \{\rho \in \text{Rep}_n R: \rho \simeq \pi\}$$

for each $\pi \in \text{Irr}_n R$ and

$$q^{-1}(q(\pi)) = \theta^{-1}(\{\rho \in \text{Rep}_\infty R: \rho \simeq \theta(\pi)\})$$

for each $\pi \in$ Fac R. □

The remaining two propositions in this section are concerned with the question of when a separable involutive Banach algebra is of type I. The propositions themselves will not be needed subsequently, although the technical lemma preceeding the second of them will be used at one point in the next section.

PROPOSITION 26.4. A separable involutive Banach algebra is of type I if and only if every factor representation of it on a separable Hilbert space is of type I.

Proof. Suppose that R is a separable involutive Banach algebra with the property that every factor representation of it on a separable Hilbert space is of type I, and consider a representation ρ of R on a Hilbert space H. If

H is separable one can show that $\rho(R)''$ is of type I by applying Proposition 19.3 and Theorem 21.2 to the central decomposition of ρ. Now suppose that H is nonseparable. By Zorn's lemma there must be a family (x_j) of unit vectors in H which is maximal with respect to the property that the (not necessarily closed) subspaces $[\pi(R) \cup \pi(R)']x_j$ are mutually orthogonal. For each j let E_j and F_j be the projections onto the closed subspaces of H spanned by $\pi(R)x_j$ and $[\pi(R) \cup \pi(R)']x_j$, respectively. Then for each j it is the case that $E_j H$ is separable (by Proposition B.1), that E_j lies in $\pi(R)'$ and that its central support in $\pi(R)'$ is F_j, hence that $\pi_{E_j}(R)'' = \pi(R)''_{E_j}$ and $\pi(R)''_{F_j}$ are *-isomorphic (by Proposition A.5), and therefore that $\pi(R)''_{F_j}$ is of type I (by what has already been proven). Moreover $\sum_j F_j = I$ by maximality, hence $\pi(R)''$ and $\oplus_j \pi(R)''_{F_j}$ are spatially isomorphic, and therefore $\pi(R)''$ is of type I. □

A Borel measure μ on $\mathrm{Irr}_\infty R$ will be called *ergodic* if every μ-measurable subset T of $\mathrm{Irr}_\infty R$ which is a union of equivalence classes is such that either T or its complement is μ-null. Notice that a Borel measure on $\mathrm{Irr}_\infty R$ which is concentrated on some equivalence class is obviously ergodic. In general, however, an ergodic Borel measure on $\mathrm{Irr}_\infty R$ need not be concentrated on an equivalence class; while this is not so easy to prove, it should be credible if it is kept in mind that Lebesgue measure on the real line is ergodic with respect to the equivalence relation under which two real numbers are equivalent if their difference is rational, yet is not concentrated by any equivalence class. Before proving the desired result relating Borel measures on $\mathrm{Irr}_\infty R$ to representations of R it will be necessary to establish the following lemma.

LEMMA 26.5. Let Z be a standard Borel space, let μ be a Borel measure on Z, let H be a Borel field of Hilbert spaces on Z, let α be a coherence for H, and let π be an α-Borel field of representations of a separable involutive Banach algebra R over H. If $\pi(\zeta)$ is irreducible for μ-a.a. $\zeta \in Z$ and if $\int_Z^\alpha \pi(\zeta)d\mu(\zeta)$ is a factor representation of type I then there is a μ-null Borel set N in Z such that the $\pi(\zeta)$, $\zeta \in Z - N$, are mutually equivalent.

Proof. Put $\rho = \int_Z^\alpha \pi(\zeta)d\mu(\zeta)$ and let M be the algebra of diagonalizable operators on $L^2(\mu;H,\alpha)$. By Theorem 13.1 and the hypothesis, there are Hilbert spaces K_1 and K_2, a linear isometry W of $L^2(\mu;H,\alpha)$ onto $K_1 \otimes K_2$, an irreducible representation σ of R on K_1, and a maximal abelian von Neumann

26. THE DUAL AND THE QUASI-DUAL

algebra N on K_2 such that $W\rho(R)"W^* = L(K_1) \otimes C(K_2)$, $W\rho(\cdot)W^* = \sigma(\cdot) \otimes I_{K_2}$, and $WMW^* = C(K_1) \otimes N$. Putting

$$W\left[\int_Z^\alpha \phi(\zeta)I_H d\mu(\zeta)\right]W^{-1} = I_{K_1} \otimes S_\phi, \quad \phi \in L^\infty(\mu)$$

one obtains a *-homomorphism $\phi \mapsto S_\phi$ of $L^\infty(\mu)$ onto N whose kernel consists of the μ-null functions in $L^\infty(\mu)$. But then one knows from Example 10.2 that there is a Borel function θ from Z to $(0,\infty)$ and a linear isometry V of K_2 onto $L^2(\theta\mu)$ such that $VS_\phi V^{-1}$ is, for each $\phi \in L^\infty(\mu) = L^\infty(\theta\mu)$, the operator on $L^2(\theta\mu)$ determined by pointwise multiplication by ϕ on $L^2(\theta\mu)$. Finally, let U be the linear isometry of $K_1 \otimes L^2(\theta\mu)$ onto $L^2(\theta\mu;K_1)$ constructed in Proposition 5.2. The operator $T = U(I \otimes V)W$ is then a linear isometry of $L^2(\mu;H,\alpha)$ onto $L^2(\theta\mu;K_1)$ and some straightforward calculations will show that TMT^{-1} is the algebra of diagonalizable operators on $L^2(\theta\mu;K_1)$ and that

$$T\left[\int_Z^\alpha \pi(\zeta)d\mu(\zeta)\right](\cdot)T^{-1} = \int_Z^\theta \sigma'(\zeta)d(\theta\mu)(\zeta)$$

where $\sigma'(\zeta) = \sigma$ for each $\zeta \in Z$. The lemma now follows from Theorem 12.4. □

PROPOSITION 26.6. Let R be a separable involutive Banach algebra. If μ is an ergodic Borel measure on $\mathrm{Irr}_\infty R$ then $\int_{\mathrm{Irr}_\infty R}^\theta \pi d\mu(\pi)$ is a factor representation which is of type I if and only if μ is concentrated on some equivalence class.

Proof. Put $Z = \mathrm{Irr}_\infty R$ and $\sigma = \int_Z^\theta \pi d\mu(\pi)$, and suppose that σ is not a factor representation. Then there will be a projection E in $\sigma(R)' \cap \sigma(R)"$ with $0 < E < I$. Now σ acts on $L^2(\mu;\ell^2)$ and by Theorem 13.1 the algebra of diagonalizable operators on $L^2(\mu;\ell^2)$ is maximal abelian in $\sigma(R)'$, and hence must contain the center of $\sigma(R)'$. In particular, there will be a Borel set S in Z with $E = \int_Z^\theta 1_S(\pi)I d\mu(\pi)$. Next, as σ is nondegenerate (Proposition 11.1) there will be a sequence (R_n) in R and a μ-null Borel set N in Z such that $\sigma(R_n) \to E$ and $\pi(R_n) \to 1_S(\pi)I$ strongly as $n \to \infty$ for each $\pi \in Z - N$ (Theorem A.2 and Proposition 6.3). The set T of all those representations in $\mathrm{Irr}_\infty R$ equivalent to some representations in $S - N$ is obviously a union of equivalence classes and, moreover, satisfies

$$S - N \subset T \subset S \cup N \tag{1}$$

Indeed, the inclusion $S - N \subset T$ is obvious from the definition of T. On the other hand, given a representation π in $T - N$, there will be a unitary operator U on ℓ_∞^2 with $U\pi(\cdot)U^* \in S - N$. But then

$$I = 1_S(U\pi(\cdot)U^*)I$$
$$= \lim_{n\to\infty} U\pi(R_n)U^*$$
$$= U[1_S(\pi)I]U^*$$
$$= 1_S(\pi)I$$

where the limit is taken in the strong operator topology, and therefore $\pi \in S$. This shows that $T - N \subset S$, or equivalently, that $T \subset S \cup N$. Now (1) implies that T is a μ-measurable subset of Z and that neither T nor $Z - T$ is μ-null, a contradiction.

This shows that σ is a factor representation. If σ is of type I then the preceeding lemma implies that μ is concentrated on an equivalence class. Conversely, if μ is concentrated on an equivalence class then σ must be of type I by Corollary 12.2. □

27. Measures on the Dual and Representations

It was mentioned in the preceeding section that the dual of a separable involutive Banach algebra is analytic (or equivalently, by Theorem 3.1 and Corollary 26.2, is countably separated) if and only if the algebra is of type I, and it is under this hypothesis that one can relate representations to measures on the dual. Accordingly, let R denote such an algebra, let p denote the canonical mapping of Irr R onto $\hat{R}$, and for each point ζ in $\hat{R}$ let $d(\zeta)$ be the unique member of $\mathbb{N} \cup \{\infty\}$ with $\zeta \in p(\text{Irr}_{d(\zeta)} R)$ and let $\gamma(\zeta)$ be the identity operator on $\ell^2_{d(\zeta)}$. Then $\ell^2_{d(\cdot)}$ is a Borel field of Hilbert spaces on $\hat{R}$ and γ is a coherence for this field. It follows rather easily from Theorem 4.3 and Corollary 26.2 that corresponding to each Borel measure μ on $\hat{R}$ there is a γ-Borel field of representations π of R over $\ell^2_{d(\cdot)}$ with $\pi(\zeta) \in \text{Irr } R$ and $p(\pi(\zeta)) = \zeta$ for μ-a.a. $\zeta \in \hat{R}$, or equivalently, with $\pi(\zeta) \in \zeta$ for μ-a.a. $\zeta \in \hat{R}$; for any such μ and π, put $\sigma(\mu,\pi) = \int_R^\gamma \pi(\zeta) d\mu(\zeta)$. Notice that by Lemma 7.2 and Theorem 12.1 the equivalence class to which $\sigma(\mu,\pi)$ belongs depends only on the measure class to which μ belongs. As the zero-representation π_0 of R on ℓ^2_1 is the only member of Irr R which fails to be nondegenerate, Proposition 11.1 implies that if R has a sequential approximate identity then $\sigma(\mu,\pi)$ is nondegenerate if and only if $\mu(\{p(\pi_0)\}) = 0$ (recall from Proposition 26.3 that $\{p(\pi_0)\}$ is a Borel set).

THEOREM 27.1. Let R be a separable involutive Banach algebra with a smooth dual. For each Borel measure μ on $\hat{R}$ and each γ-Borel field of representations

27. MEASURES ON THE DUAL AND REPRESENTATIONS

π of R over $\ell^2_{d(\cdot)}$ with $\pi(\zeta) \in \zeta$ for μ-a.a. $\zeta \in \hat{R}$ the representation $\sigma(\mu,\pi)$ is multiplicity-free and, conversely, each multiplicity-free representation of R on a separable Hilbert space is equivalent to one of this form.

Proof. Suppose that μ is a Borel measure on $\hat{R}$ and that π is a γ-Borel field of representations of R over $\ell^2_{d(\cdot)}$ with $\pi(\zeta) \in \zeta$ for μ-a.a. $\zeta \in \hat{R}$. The algebra M of diagonalizable operators on $L^2(\mu; \ell^2_{d(\cdot)}, \gamma)$ is a maximal abelian subalgebra of $\sigma(\mu,\pi)(R)'$ by Theorem 13.1 and hence contains the center N of $\sigma(\mu,\pi)(R)'$. So to show that $N = \sigma(\mu,\pi)(R)'$, i.e., that $\sigma(\mu,\pi)$ is multiplicity-free, it is evidently sufficient to show that $M = N$.

Notice that by Lemma 7.2 one may assume that μ is finite. Putting $X = \hat{R}$ and $\alpha = \gamma$ and letting $Y, \nu, \omega, (\mu_\xi)_{\xi \in Y}$, and β be as in Example 10.9, one knows from Example 15.4 that $\xi \mapsto \int_X^\gamma \pi(\zeta) d\mu_\xi(\zeta)$ is a β-Borel field of representations of R and that $\int_Y^\beta \int_X^\gamma \pi(\zeta) d\mu_\xi(\zeta) d\nu(\xi)$ is a direct integral decomposition of $\sigma(\mu,\pi)$ with respect to N. Let M be a μ-null Borel subset of X such that $\pi(\zeta) \in \mathrm{Irr}\, R$ and $p(\pi(\zeta)) \in \zeta$ for all $\zeta \in X - M$. Then Theorem 13.2 (b) and the formula in part (b) of Theorem 4.5 imply that there is a ν-null Borel subset P of Y such that $\int_X^\gamma \pi(\zeta) d\mu_\xi(\zeta)$ is a factor representation and such that $\mu_\xi(M) = 0$ for each $\xi \in Y - P$. But then Lemma 26.5 implies that for each point ξ in $Y - P$ there is a μ_ξ-null Borel set M_ξ in X such that the representations $\pi(\zeta)$, $\zeta \in X - M_\xi$, are mutually equivalent.

Now let S be some given Borel set in X and put $T = \{\xi \in Y : \mu_\xi(S) > 0\}$. If ξ is a point in $Y - P$ for which $\mu_\xi(X - M) > 0$ then $X - (M \cup M_\xi)$ is a Borel subset of X which is not μ_ξ-null and which can contain at most one point, and hence $X - (M \cup M_\xi) = \{\zeta\}$ for some point ζ in X which must necessarily lie in $\omega^{-1}(\xi)$ and not in $S \Delta \omega^{-1}(T)$. Indeed, $\zeta \in \omega^{-1}(\xi)$ as $\mu_\xi(X - \omega^{-1}(\xi)) = 0$, and $\zeta \notin S \Delta \omega^{-1}(T)$ as

$$\zeta \in \omega^{-1}(T) \iff \mu_\xi(S) > 0$$
$$\iff \mu_\xi(S - (M \cup M_\xi)) > 0$$
$$\iff \zeta \in S$$

This proves that

$$\mu_\xi((S \Delta \omega^{-1}(T)) \cap (X - M)) = 0$$

for all $\xi \in Y - P$, and hence that

$$\mu(S \triangle \omega^{-1}(T)) = \mu((S \triangle \omega^{-1}(T)) \cap (X - M))$$
$$= \int_Y \mu_\xi ((S \triangle \omega^{-1}(T)) \cap (X - M)) d\nu(\xi)$$
$$= 0$$

or equivalently, that $1_S = 1_T \circ \omega$ μ-a.e. From this it follows easily that the mapping of $L^\infty(\nu)$ into $L^\infty(\mu)$ induced by the mapping $\psi \mapsto \psi \circ \omega$ of $L^\infty(\nu)$ into $L^\infty(\mu)$ is actually onto, and therefore that $M = N$.

Now suppose that, conversely, one is given a multiplicity-free representation σ of R on a separable Hilbert space. Then from Theorems 12.3, 13.1, and 13.3 one knows that there is a standard Borel space Z, a finite Borel measure ν on Z, a Borel field of Hilbert spaces H on Z, a coherence α for H, and an α-Borel field of representations ρ of R over H such that $\int_Z^\alpha \rho(\zeta) d\nu(\zeta)$ is the central decomposition of σ and such that the representations $\rho(\zeta)$, $\zeta \in Z$, are irreducible and mutually inequivalent. Moreover, one may clearly assume that $\dim H(\zeta) > 0$, that $H(\zeta) = \ell^2_{\dim H(\zeta)}$, and that $\alpha(\zeta) = I$ for each $\zeta \in Z$. Then ρ is actually a Borel mapping of Z into Irr R such that $p \circ \rho$ is one-one. Put $\mu = (p \circ \rho)_*(\nu)$, and let $\pi(\zeta) = \rho((p \circ \rho)^{-1}(\zeta))$ for $\zeta \in \text{range}(p \circ \rho)$ and let $\pi(\zeta)$ be the zero-representation of R on $\ell^2_{d(\zeta)}$ for $\zeta \in \hat{R} - \text{range}(p \circ \rho)$. Then it follows from Corollary 2.12 and Theorem 2.13 that π is a γ-Borel field of representations of R over $\ell^2_{d(\cdot)}$ with $\pi(\zeta) \in \zeta$ for μ-a.a. $\zeta \in \hat{R}$, and it is clear that σ and $\sigma(\mu,\pi)$ are equivalent. □

THEOREM 27.2. Let R be a separable involutive Banach algebra with a smooth dual. Two Borel measures μ and ν on $\hat{R}$ are equivalent [respectively, singular, satisfy $\mu \ll \nu$] if and only if for any two γ-Borel fields of representations π and ρ of R over $\ell^2_{d(\cdot)}$ with $\pi(\zeta) \in \zeta$ for μ-a.a. $\zeta \in \hat{R}$ and $\rho(\zeta) \in \zeta$ for ν-a.a. $\zeta \in \hat{R}$ the representations $\sigma(\mu,\pi)$ and $\sigma(\nu,\rho)$ are equivalent [respectively, disjoint, satisfy $\sigma(\mu,\pi) \leq \sigma(\nu,\rho)$].

Proof. Let μ, ν, π, and ρ be as in the statement of the theorem. It has already been pointed out that if μ and ν are equivalent then so are $\sigma(\mu,\pi)$ and $\sigma(\nu,\rho)$; conversely, if $\sigma(\mu,\pi)$ and $\sigma(\nu,\rho)$ are equivalent then it follows without any difficulty from Theorems 9.2, 12.4, and 27.1 that μ and ν must be equivalent.

If $\sigma(\mu,\pi) \leq \sigma(\nu,\rho)$ then, by definition, $\sigma(\mu,\pi) \simeq \sigma(\nu,\rho)_E$ for some projection E in $\sigma(\nu,\rho)(R)'$. Now E must be of the form $\int_{\hat{R}}^\gamma 1_S(\zeta) I d\nu(\zeta)$ for some Borel set S in $\hat{R}$ (by Theorems 27.1 and 13.1), and it is not hard to see that $\sigma(\nu,\rho)_E \simeq \sigma(1_S \nu, \rho)$. But then $\mu \ll \nu$ by what was just proven. Con-

27. MEASURES ON THE DUAL AND REPRESENTATIONS 123

versely, if $\mu \ll \nu$ then $\mu = f\nu$ for some nonnegative-valued Borel function f on $\hat{R}$ by the Radon-Nikodym theorem. Put $S = \{\zeta \in \hat{R}: f(\zeta) \neq 0\}$ and $E = \int_{\hat{R}}^{\oplus} 1_S(\zeta) \mathrm{Id}\nu(\zeta)$. Then μ and $1_S\nu$ are equivalent and so $\sigma(\mu,\pi)$ and $\sigma(1_S\nu,\rho)$ are equivalent. But as was just pointed out, $\sigma(1_S\nu,\rho)$ and $\sigma(\nu,\rho)_E$ are also equivalent. This shows that $\sigma(\mu,\pi) \leq \sigma(\nu,\rho)$.

If $\sigma(\mu,\pi)$ and $\sigma(\nu,\rho)$ fail to be disjoint then by Proposition B.3 there will be nonzero projections E and F in $\sigma(\mu,\pi)(R)'$ and $\sigma(\nu,\rho)(R)'$, respectively, such that the representations $\sigma(\mu,\pi)_E$ and $\sigma(\nu,\rho)_F$ are equivalent. The argument in the preceeding paragraph then shows that there will be Borel subsets S and T of $\hat{R}$ such that $\mu(S) > 0$, $\nu(T) > 0$, and $\sigma(1_S\mu,\pi) \simeq \sigma(1_T\nu,\rho)$. But then $1_S\mu$ and $1_T\nu$ will be equivalent by what was already proven, and so μ and ν cannot be mutually singular. Conversely, if one assumes that μ and ν are not mutually singular then there will be a Borel subset S of $\hat{R}$ such that the measures $1_S\mu$ and $1_S\nu$ are nonzero and equivalent. Indeed, applying the Lebesgue decomposition theorem twice gives Borel sets M and N in $\hat{R}$ satisfying

$$1_M\mu \ll \nu \quad \text{and} \quad 1_{\hat{R}-M} \perp \nu \tag{1}$$

and

$$1_N\nu \ll 1_M\mu \quad \text{and} \quad 1_{\hat{R}-N}\nu \perp 1_M\mu \tag{2}$$

here $\perp$ denotes mutual singularity of measures. From these relations it is clear that the measures $1_{M \cap N}\mu$ and $1_{M \cap N}\nu$ are equivalent. If $1_{M \cap N}\mu = 0$ then $1_N\nu = 0$ and $\nu \perp 1_M\mu$ by (2), and hence $\mu \perp \nu$ by (1). Thus $S = M \cap N$ will do. But then $\sigma(1_S\mu,\pi) \simeq \sigma(1_S\nu,\rho)$ by what has already been proven, and as $\sigma(1_S\mu,\pi) \leq \sigma(\mu,\pi)$ and $\sigma(1_T\nu,\rho) \leq \sigma(\nu,\rho)$ by a previous remark, it follows from Proposition B.3 that $\sigma(\mu,\pi)$ and $\sigma(\nu,\rho)$ cannot be disjoint. This completes the proof of Theorem 27.2. □

The two theorems just proven can be both summarized and put into a more elegant form by passing to classes of representations and measures. Namely, the theorems assert that there is a one-one correspondence between Borel measure classes (i.e., equivalence classes of Borel measures) on $\hat{R}$ and equivalence classes of multiplicity-free representations of R on separable Hilbert spaces which respects

(i) the partial orderings "$\leq$" on classes of representations and "$\ll$" on Borel measure classes, and

(ii) the relations "$\wr$" on classes of representations and "$\perp$" on Borel measure classes.

In order to bring all representations of R on separable Hilbert spaces into this picture it should be pointed out that there is a one-one correspondence between equivalence classes of multiplicity-free representations of R on separable Hilbert spaces and quasi-equivalence classes of representations of R on separable Hilbert spaces. Indeed, by Proposition A.3 and Theorem A.6 each quasi-equivalence class of representations of R on separable Hilbert spaces contains exactly one equivalence class of multiplicity-free representations of R on separable Hilbert spaces.

It will be worthwhile to conclude this section with a detailed consideration of a construction which associates with any representation on a separable Hilbert space a quasi-equivalent multiplicity-free representation. This construction is analogous to part of the proof of Lemma 24.4 and can be interpreted to mean that one passes from a given representation to a quasi-equivalent multiplicity-free one by "removing all multiplicities" or "making all nonzero multiplicities equal to one".

Let R continue to denote a separable involutive Banach algebra of type I, and consider the direct integral decomposition of a representation π of R on a separable Hilbert space with respect to a maximal abelian subalgebra of $\pi(R)'$. In this way one obtains (by Theorem 13.1) a standard Borel space Z, a finite Borel measure μ on Z, and a Borel map ρ from Z into Irr R such that if, for each point ζ in Z, $\alpha(\zeta)$ denotes the identity operator on the Hilbert space on which $\rho(\zeta)$ acts then $\pi \simeq \int_Z^\alpha \rho(\zeta) d\mu(\zeta)$. Let p denote the canonical mapping of Irr R onto $\hat{R}$ and put $(p \circ \rho)_*(\mu) = \nu$. Now as R is of type I (or equivalently, has a smooth dual) the range Y of $p \circ \rho$ will be an analytic subset of $\hat{R}$. But then (by Theorems 4.1 and 4.3 and the fact that one can in effect delete a given ν-null Borel set from $\hat{R}$ by deleting a suitable μ-null Borel set from Z) one may assume that Y is a Borel set and that there is a Borel function σ from Y to Z with $(p \circ \rho \circ \sigma)(\xi) = \xi$ for $\xi \in Y$. Letting $\sigma(\xi)$ be some fixed point in Z for each point ξ in $\hat{R} - Y$, it will then be the case that $p \circ \sigma$ is a γ-Borel field of representations of R over $\ell^2_{d(\cdot)}$ satisfying $(p \circ \sigma)(\xi) \in$ Irr R and $p((\rho \circ \sigma)(\xi)) = \xi$ for ν-a.a. $\xi \in \hat{R}$, and hence one can form the representation $\sigma(\nu, \rho \circ \sigma)$. Next, one may regard $p \circ \rho$ as a map from Z onto Y and then apply Theorem 4.5 to find a Borel map $\xi \mapsto \mu_\xi$ from Y to $M(Z)$ satisfying the conclusions of that theorem. If $n(\zeta)$ denotes the dimension of the Hilbert space on which $\rho(\zeta)$

27. MEASURES ON THE DUAL AND REPRESENTATIONS

acts for each point $\zeta \in Z$ and if ν_Y denotes the restriction of ν to the Borel subsets of Y then Lemma 24.1 implies that $\xi \mapsto L^2(\mu_\xi; \ell^2_{n(\cdot)}, \alpha)$ is a Borel field of Hilbert spaces on Y and that there is a coherence β for this field such that $\xi \mapsto \int_Z^\alpha \rho(\zeta) d\mu_\xi(\zeta)$ is a β-Borel field of representations and such that

$$\pi \simeq \int_Y^\beta \int_Z^\alpha \rho(\zeta) d\mu_\xi(\zeta) d\nu_Y(\xi) \tag{1}$$

For each point ξ in Y it is the case $\rho(\zeta) \simeq (\rho \circ \sigma)(\xi)$ for μ_ξ-a.a. $\zeta \in Z$, and hence, if I_ξ denotes the identity operator on $L^2(\mu_\xi)$, that

$$\int_Z^\alpha \rho(\zeta) d\mu_\xi(\zeta) \simeq (\rho \circ \sigma)(\xi)(\cdot) \otimes I_\xi \tag{2}$$

by Corollary 12.2.

In view of the last two equations the next step is evidently to form the direct integral of the field of representations $\xi \mapsto (\rho \circ \sigma)(\xi)(\cdot) \otimes I_\xi$ on Y, and to do this one needs a suitable coherence. One way to construct such a coherence is to proceed as follows: Put $m(\xi) = \dim L^2(\mu_\xi)$ for each point ξ in Y and let $\theta_{p,q}$ be some fixed linear isometry of $\ell^2_p \otimes \ell^2_q$ onto ℓ^2_{pq} for each pair of elements p and q in $\mathbb{N} \cup \{\infty\}$. From Lemma 24.1 applied to the constant field of Hilbert spaces $\xi \mapsto \mathbb{C}$ on Y one knows that $\xi \mapsto L^2(\mu_\xi)$ is a Borel field of Hilbert spaces on Y and that there is a coherence δ for this field with the property that $\xi \mapsto f(\mu_\xi)$ is a δ-Borel vector field for any bounded complex-valued Borel function f on Z. Thus $\xi \mapsto \ell^2_{n(\sigma(\xi))} \otimes L^2(\mu_\xi)$ is a Borel field of Hilbert spaces on Y and one can define a coherence ε for this field by putting

$$\varepsilon(\xi) = \theta_{n(\sigma(\xi)), m(\xi)} \circ [I \otimes \delta(\xi)], \quad \xi \in Y$$

To show that ε is a suitable coherence, i.e., that the field of representations $\xi \mapsto (\rho \circ \sigma)(\xi)(\cdot) \otimes I_\xi$ is ε-Borel, it is sufficient to show that

$$\xi \mapsto \varepsilon(\xi)[(\rho \circ \sigma)(\xi)(R) \otimes I_\xi][v(\xi) \otimes f(\mu_\xi)]$$

is a Borel map from Y to ℓ^2 for any R in R, any Borel function v from Y to ℓ^2 with $v(\xi) \in \ell^2_{n(\sigma(\xi))}$ for each $\xi \in Y$, and any bounded complex-valued Borel function f on Z. Now Y is the union of the mutually disjoint Borel sets $(n \circ \sigma)^{-1}(p) \cap m^{-1}(q)$, $p,q \in \mathbb{N} \cup \{\infty\}$, and the value of the function in question at a point ξ in $(n \circ \sigma)^{-1}(p) \cap m^{-1}(q)$ is

$$\theta_{p,q}[(\rho \circ \sigma)(\xi)(R)v(\xi) \otimes \delta(\xi)f(\mu_\xi)]$$

and hence the function is Borel.

It now follows from formulae (1) and (2) and Theorem 12.1 that

$$\pi \simeq \int_Y^\oplus (\rho \circ \sigma)(\xi)(\cdot) \otimes I_\xi d\nu_Y(\xi)$$

$$\approx \int_{\hat{R}}^\oplus (\rho \circ \sigma)(\xi) d\nu(\xi)$$

$$= \sigma(\nu, \rho \circ \sigma)$$

It is clear from this calculation, Theorem 27.2, and Proposition A.3 that the Borel measure class to which ν belongs depends only on the quasi-equivalence class to which π belongs (and, in particular, not on the specific maximal decomposition under consideration), and hence that the equivalence classes of the irreducible representations which appear as the components of a maximal direct integral decomposition of π are essentially uniquely determined (i.e., determined up to a ν-null set). This latter result is, of course, in sharp contrast to the non-type I situation (cf. Example 15.2).

28. Measures on the Quasi-dual and Representations

Consider a separable involutive Banach algebra R. Assuming for the moment that R has a smooth dual (or equivalently, is of type I), the previous section contains a canonical procedure for analyzing a multiplicity-free representation ρ of R on a separable Hilbert space. Namely, the Borel space which appears in the direct integral decomposition of ρ with respect to $\rho(R)'$ can be taken to be the dual $\hat{R}$ of R, and in this way ρ determines a Borel measure on $\hat{R}$ whose properties turn out to be closely related to those of ρ. Notice that the decomposition which is used here is both central and maximal. Recall also that one can obtain some information about an arbitrary representation of R on a separable Hilbert space as any such representation is quasi-equivalent to a multiplicity-free representation of R on a separable Hilbert space and that this latter representation is unique up to equivalence. This analysis quite clearly breaks down completely if R fails to be of type I, and it is tempting to search for an analogous procedure for studying representation of R on separable Hilbert spaces, and preferably one

28. MEASURES ON THE QUASI-DUAL AND REPRESENTATIONS

which reduces to the one just described in case R is of type I and the representation in question is multiplicity-free. Such a procedure does in fact exist but is technically somewhat more complicated that the one discussed in Section 27 and will only be described after several preliminary results have been established. Notice that the first of these is a strengthening of part of Theorem 12.1. Before even turning to these results it will, however, be useful to motivate the procedure with some simple remarks.

Consider a representation π of R on a separable Hilbert space. A reasonable first step in trying to construct a procedure for studying π would be to form in some canonical manner a direct integral decomposition of π, or what is essentially equivalent, to select in a canonical manner an abelian von Neumann subalgebra of $\pi(R)'$. Once this is admitted it can be argued that one has no choice but to form the central decomposition. It is certainly canonical; the only other possible choice would seem to be some maximal decomposition, but this has to be rejected since the various maximal decompositions of π may be completely unrelated to one another (cf. Example 15.2) and since there is no canonical method for selecting one of them. Moreover, because of Corollary 13.4 it seems reasonable to expect to be able to take the quasi-dual $\widehat{R}$ of R as the Borel space appearing in the central decomposition of π, hence to associate a Borel measure class on $\widehat{R}$ to π in a canonical manner, and then to relate properties of π to those of this measure class much like what was done in Section 27.

LEMMA 28.1. Let R be a separable involutive Banach algebra, let Z be a standard Borel space, let μ be a Borel measure on Z, and for $j = 1,2$, let H_j be a Borel field of Hilbert spaces on Z, let α_j be a coherence for H_j, let π_j be an α_j-Borel field of representations of R over H_j, let M_j be the algebra of diagonalizable operators on $L^2(\mu; H_j, \alpha_j)$, and put $\rho_j = \int_Z^{\alpha_j} \pi_j(\zeta) d\mu(\zeta)$. If $\pi_1(\zeta)$ and $\pi_2(\zeta)$ are quasi-equivalent for μ-a.a. $\zeta \in Z$ then there is a *-isomorphism of $[\rho_1(R) \cup M_1]''$ onto $[\rho_2(R) \cup M_2]''$ carrying $\rho_1(R)$ onto $\rho_2(R)$, $R \in R$, and M_1 onto M_2.

Proof. Put $K_j(\zeta) = H_j(\zeta) \otimes \ell^2$ and $\sigma_j(\zeta)(R) = \pi_j(\zeta)(R) \otimes I$ for $j = 1,2$, for $\zeta \in Z$, and for $R \in R$. It is apparent from Lemma 7.4 that for $j = 1,2$ there is a coherence β_j for the field of Hilbert spaces K_j on Z and a linear isometry W_j of $L^2(\mu; K_j, \beta_j)$ onto $L^2(\mu; H_j, \alpha_j) \otimes \ell^2$ such that σ_j is a β_j-Borel field of representations of R over K_j, such that W_j intertwines $\int_Z^{\beta_j} \sigma_j(\zeta) d\mu(\zeta)$ and $\rho_j(\cdot) \otimes I$, and such that $W_j^*[M_j \otimes C(\ell^2)]W_j$ is the algebra of diagonalizable operators on $L^2(\mu;K_j,\beta_j)$. Now $\sigma_1(\zeta)$ and $\sigma_2(\zeta)$ are equivalent for μ-a.a. $\zeta \in Z$ by Proposition A.3, and so from Theorem 12.1 and its proof one knows that there is a linear isometry W of $L^2(\mu;K_1,\beta_1)$ onto $L^2(\mu;K_2,\beta_2)$ intertwining $\int_Z^{\beta_1} \sigma_1(\zeta)d\mu(\zeta)$ and $\int_Z^{\beta_2} \sigma_2(\zeta)d\mu(\zeta)$ and satisfying

$$WW_1^*[M_1 \otimes C(\ell^2)]W_1W^* = W_2^*[M_2 \otimes C(\ell^2)]W_2$$

Thus $V = W_2WW_1^*$ is a linear isometry of $L^2(\mu;H_1,\alpha_1) \otimes \ell^2$ onto $L^2(\mu;H_2,\alpha_2) \otimes \ell^2$ intertwining the representations $\rho_1(\cdot) \otimes I$ and $\rho_2(\cdot) \otimes I$ and satisfying

$$V[M_1 \otimes C(\ell^2)]V^* = M_2 \otimes C(\ell^2)$$

It should now be clear that the desired *-isomorphism can be constructed from V and the usual *-isomorphisms between $L(L^2(\mu;H_j,\alpha_j))$ and $L(L^2(\mu;H_j,\alpha_j)) \otimes C(\ell^2)$, $j = 1,2$. □

For any two representations π and ρ of an involutive Banach algebra R let $I(\pi,\rho)$ be the dimension of the vector space of operators which intertwine π and ρ if this dimension is finite, and otherwise put $I(\pi,\rho) = \infty$. In this notation one can paraphrase Proposition B.5 by saying that if π and ρ are both factorial then $I(\pi,\rho) \geq 1$ if and only if π and ρ are quasi-equivalent.

LEMMA 28.2. For a separable involutive Banach algebra R the map $(\pi,\rho) \mapsto I(\pi,\rho)$ from $(\text{Rep}_\infty R) \times (\text{Rep}_\infty R)$ to $\{0\} \cup \mathbb{N} \cup \{\infty\}$ is Borel.

Proof. The proof of this lemma is very similar to that of Lemma 17.8. Let (R_n) be a dense sequence in the unit ball of R, let $L = L(\ell^2)$, and let L_*, M, and M_* be as in the proof of Lemma 17.8. For each pair of representations π and ρ of R on ℓ^2 let $S_{\pi,\rho}$ denote the linear mapping from M_* to L^* defined by

$$[S_{\pi,\rho}((f_n))](A) = \sum_{n=1}^{\infty} f_n(A\pi(R_n) - \rho(R_n)A)$$

where $(f_n) \in M_*$ and $A \in L$ are arbitrary. Then (just as in the proof of Lemma 17.8) one can show that $S_{\pi,\rho}$ is actually a norm-continuous linear operator from M_* to L_*. The dual $S_{\pi,\rho}^*$ of $S_{\pi,\rho}$ is then the linear map from L to M given by

$$(S_{\pi,\rho}^*(A))_n = A\pi(R_n) - \rho(R_n)A, \qquad A \in L$$

28. MEASURES ON THE QUASI-DUAL AND REPRESENTATIONS

Now it is clear from Proposition B.1 that

$$\dim \ker S_{\pi,\rho}^* = I(\pi,\rho)$$

and hence from the general theory of linear operators on Banach spaces that

$$\operatorname{codim}[S_{\pi,\rho}(M_*)]^- = I(\pi,\rho)$$

Let (h_k) be a dense sequence in M_*. Suppose that $f_1, \ldots, f_m$ are m linearly independent elements of L_*, where m is some fixed positive integer, and let D be the set of all m-tuples of complex numbers $(r_1, \ldots, r_m)$ with $r_j \in Q + iQ$ for each j and with $1 \leq |r_1|^2 + \ldots + |r_m|^2 \leq 2$. Then for all m-tuples $r = (r_1, \ldots, r_m)$ in D and all pairs of positive integers k and p the subset $S(r;k,p)$ of $(\operatorname{Rep}_\infty R) \times (\operatorname{Rep}_\infty R)$ consisting of all those points (π,ρ) for which

$$\| r_1 f_1 + \cdots + r_m f_m - S_{\pi,\rho}(h_k) \| < p^{-1}$$

is Borel. One can now show without much difficulty that the $f_1, \ldots, f_m$ are linearly dependent modulo $[S_{\pi,\rho}(M_*)]^-$ if and only if for each $p \in \mathbb{N}$ there is an $r \in D$ and a $k \in \mathbb{N}$ such that $(\pi,\rho) \in S(r;k,p)$, i.e., if and only if

$$(\pi,\rho) \in \cap_{p=1}^\infty \cup_{k=1}^\infty \cup_{r \in D} S(r;k,p)$$

This shows that the subset of $(\operatorname{Rep}_\infty R) \times (\operatorname{Rep}_\infty R)$ consisting of those pairs (π,ρ) for which $f_1, \ldots, f_m$ are linearly dependent modulo $[S_{\pi,\rho}(M_*)]^-$ is Borel. Now let (g_j) be a linearly independent and total sequence in L_*. Then for any $n \in \{0\} \cup \mathbb{N}$ the codimension of $[S_{\pi,\rho}(M_*)]^-$ in L_* is at most n if and only if every finite subset of $\{g_j : j \in \mathbb{N}\}$ with more than n elements is linearly dependent modulo $[S_{\pi,\rho}(M_*)]^-$; and as there are but countably many such subsets it follows that the set

$$\{(\pi,\rho) \in (\operatorname{Rep}_\infty R) \times (\operatorname{Rep}_\infty R) : I(\pi,\rho) \leq n\}$$

is Borel. But this obviously implies that I is a Borel function from $(\operatorname{Rep}_\infty R) \times (\operatorname{Rep}_\infty R)$ to $\{0\} \cup \mathbb{N} \cup \{\infty\}$. □

PROPOSITION 28.3. Let R be a separable involutive Banach algebra. If S is a Borel subset of Fac R consisting of mutually nonquasi-equivalent representations then the set

$\{\rho \in \text{Fac } R: \rho \approx \pi \text{ for some } \pi \in S\}$

is Borel.

Proof. Let T denote the subset of (Fac R) × (Fac R) consisting of those pairs (π,ρ) for which $\pi \approx \rho$. If θ is the Borel map from Fac R to Fac$_\infty R$ constructed in the proof of Proposition 26.3 then by Proposition B.5 a pair (π,ρ) in (Fac R) × (Fac R) belongs to T if and only if $I(\theta(\pi),\theta(\rho)) \geq 1$. This observation together with Lemma 28.2 shows that T, and hence also U = T ∩ (Fac R × S), is a Borel subset of (Fac R) × (Fac R). Now the restriction to U of the projection p of (Fac R) × (Fac R) onto its first coordinate is one-one by assumption, and therefore p(U) is a Borel subset of Fac R by Theorem 2.13. But p(U) is the set in question, i.e.,

$p(U) = \{\rho \in \text{Fac } R: \rho \approx \pi \text{ for some } \pi \in S\}$ □

The preliminary results now having been disposed of, one can begin the construction which will associate a Borel measure on the quasi-dual of a separable involutive Banach algebra to each representation of that algebra on a separable Hilbert space. Accordingly, let R be a separable involutive Banach algebra and let q denote the canonical mapping of Fac R onto $\widehat{R}$. Suppose that one is given a standard Borel measure μ on $\widehat{R}$. Then it follows easily from Theorem 4.3 and Corollary 26.2 that there will be a Borel map π from $\widehat{R}$ to Fac R satisfying $q(\pi(\zeta)) = \zeta$ for μ-a.a. $\zeta \in \widehat{R}$, or equivalently, satisfying $\pi(\zeta) \in \zeta$ for μ-a.a. $\zeta \in \widehat{R}$. If $d(\zeta)$ denotes the dimension of the Hilbert space on which $\pi(\zeta)$ acts and $\alpha(\zeta)$ the identity operator on $\ell^2_{d(\zeta)}$ for each $\zeta \in \widehat{R}$ then $\ell^2_{d(\cdot)}$ is a Borel field of Hilbert spaces on $\widehat{R}$, α is a coherence for this field, and π is an α-Borel field of representations of R over $\ell^2_{d(\cdot)}$. One can therefore form the direct integral representation $\int_{\widehat{R}}^{\alpha} \pi(\zeta) d\mu(\zeta)$, and as α is determined by π one can unambiguously denote this representation by $\tau(\mu,\pi)$. Notice that by Lemmas 7.2 and 28.1 the quasi-equivalence class of representations of R to which $\tau(\mu,\pi)$ belongs and whether or not $\tau(\mu,\pi)$ is a central decomposition depends only on the measure class to which μ belongs. If $\tau(\mu,\pi)$ is a central decomposition, i.e., if the center of $\tau(\mu,\pi)(R)''$ is the algebra of diagonalizable operators on $L^2(\mu;\ell^2_{d(\cdot)},\alpha)$, then both μ and the measure class to which it belongs will be called *canonical*. An example will be given later to show that in general not every standard Borel measure on $\widehat{R}$ is canonical.

28. MEASURES ON THE QUASI-DUAL AND REPRESENTATIONS

THEOREM 28.4. For each representation σ of R on a separable Hilbert space there is a unique Borel measure class on $\widehat{R}$ with the property that if μ is any member of this measure class then μ is standard and canonical and there is a Borel map π from $\widehat{R}$ to Fac R with $\pi(\zeta) \in \zeta$ for μ-a.a. $\zeta \in \widehat{R}$ and such that $\tau(\mu,\pi)$ is a central decomposition of σ.

Proof. The proof of uniqueness will be taken up first. Suppose that μ and ν are two standard and canonical Borel measures on $\widehat{R}$, that π and ρ are Borel maps from $\widehat{R}$ to Fac R satisfying $\pi(\zeta) \in \zeta$ for μ-a.a. $\zeta \in \widehat{R}$ and $\rho(\zeta) \in \zeta$ for ν-a.a. $\zeta \in \widehat{R}$, and that $\tau(\mu,\pi)$ and $\tau(\nu,\rho)$ are equivalent. In view of Lemma 7.2 one may as well assume that both μ and ν are finite. As μ and ν are canonical any linear isometry from the Hilbert space on which $\tau(\mu,\pi)$ acts onto that on which $\tau(\rho,\nu)$ acts which intertwines $\tau(\mu,\pi)$ and $\tau(\nu,\rho)$ will induce a *-isomorphism between the corresponding algebras of diagonalizable operators, and hence (as μ and ν are standard) one can apply Theorems 9.2 and 12.4. According to these two theorems, there is a Borel map ω from $\widehat{R}$ to itself such that $\omega_*(\mu)$ and ν are equivalent and there are Borel subsets M and N of $\widehat{R}$ satisfying the following conditions: $\mu(M) = \nu(N) = 0$; the restriction of ω to $\widehat{R} - M$ is one-one; $q(\pi(\zeta)) = \zeta$ if $\zeta \in \widehat{R} - M$ and $q(\rho(\zeta)) = \zeta$ if $\zeta \in \widehat{R} - N$; and $\rho(\omega(\zeta)) \simeq \pi(\zeta)$ if $\zeta \in \widehat{R} - M$. Let P denote the complement of $M \cup \omega^{-1}(N)$ in $\widehat{R}$ and notice that $\mu(\widehat{R} - P) = 0$ and that for any point ζ in P one has

$$\omega(\zeta) = q(\rho(\omega(\zeta))) = q(\pi(\zeta)) = \zeta$$

From this it follows that for any Borel set S in $\widehat{R}$ one has $S \cap P = \omega^{-1}(S) \cap P$, and hence

$$\mu(S) = 0 \iff \mu(S \cap P) = 0$$
$$\iff \mu(\omega^{-1}(S) \cap P) = 0$$
$$\iff \mu(\omega^{-1}(S)) = 0$$
$$\iff \nu(S) = 0$$

Turning now to the existence part of the theorem, say that σ is a representation of R on a separable Hilbert space. One knows from Theorem 12.3 and Corollary 13.4 that there is a standard Borel space Z, a finite Borel measure ν on Z, a Borel field of Hilbert spaces H on Z, a coherence α for H, and an α-Borel field of representations ρ of R over H such that the $\rho(\zeta)$,

$\zeta \in Z$, are mutually disjoint factor representations and such that $\int_Z^\alpha \rho(\zeta)d\nu(\zeta)$ is a central decomposition of σ. Moreover, one may clearly assume that dim $H(\zeta) > 0$, that $H(\zeta) = \ell^2_{\dim H(\zeta)}$ and that $\alpha(\zeta) = I$ for each $\zeta \in Z$. Then ρ is actually a Borel mapping of Z into Fac R with the property that $q \circ \rho$ is one-one. The range W of $q \circ \rho$ is a Borel subset of $\widehat{R}$ and $q \circ \rho$ is a Borel isomorphism of Z onto W, where W is given the relative Borel structure. Indeed, if S is a Borel subset of Z then

$$q^{-1}((q \circ \rho)(S)) = \{\pi \in \text{Fac } R: \pi \approx \pi' \text{ for some } \pi' \in \rho(S)\}$$

is a Borel set by Theorem 2.13 and Proposition 28.3, and thus $(q \circ \rho)(S)$ is a Borel subset of $\widehat{R}$. This means that $\mu = (q \circ \rho)_*(\nu)$ is a standard Borel measure on $\widehat{R}$ and that if one puts $\pi(\zeta) = \rho((q \circ \rho)^{-1}(\zeta))$ for $\zeta \in W$ and lets $\pi(\zeta)$ be the zero-representation of R on ℓ^2 for $\zeta \in \widehat{R} - W$ then π is a Borel map from $\widehat{R}$ to Fac R satisfying $\pi(\zeta) \in \zeta$ for μ-a.a. $\zeta \in \widehat{R}$. It should now be clear that $\tau(\mu,\pi)$ and $\int_Z^\alpha \rho(\zeta)d\nu(\zeta)$ are equivalent, that $\tau(\mu,\pi)$ is a central decomposition of σ, and hence that μ is canonical. □

THEOREM 28.5. Suppose that μ and ν are two standard and canonical Borel measures on $\widehat{R}$ and that π and ρ are two Borel maps from $\widehat{R}$ to Fac R satisfying $\pi(\zeta) \in \zeta$ for μ-a.a. $\zeta \in \widehat{R}$ and $\rho(\zeta) \in \zeta$ for ν-a.a. $\zeta \in \widehat{R}$. Then μ and ν are equivalent [respectively, singular] if and only if $\tau(\mu,\pi)$ and $\tau(\nu,\rho)$ are quasi-equivalent [respectively, disjoint], and $\mu \ll \nu$ if and only if $\tau(\mu,\pi)$ is quasi-equivalent to a subrepresentation $\tau(\nu,\rho)$.

Proof. Let μ,ν,π, and ρ be as in the statement of the theorem. If μ and ν are equivalent then $\tau(\mu,\pi)$ and $\tau(\nu,\rho)$ are quasi-equivalent by Lemma 7.2 and Theorem 12.1. Now suppose that, conversely, $\tau(\mu,\pi)$ and $\tau(\nu,\rho)$ are quasi-equivalent. For $n \in \mathbb{N} \cup \{\infty\}$ let W_n be a linear isometry of $\ell^2_n \otimes \ell^2$ onto ℓ^2. For each point ζ in $\widehat{R}$ let $m(\zeta)$ and $n(\zeta)$ be the dimensions of the Hilbert spaces on which $\pi(\zeta)$ and $\rho(\zeta)$, respectively, act, and let

$$\pi'(\zeta) = W_{m(\zeta)}[\pi(\zeta)(\cdot) \otimes I]W_{m(\zeta)}^*$$

and

$$\rho'(\zeta) = W_{n(\zeta)}[\rho(\zeta)(\cdot) \otimes I]W_{n(\zeta)}^*$$

Then π' and ρ' are Borel functions from $\widehat{R}$ to $\text{Fac}_\infty R$ satisfying $\pi'(\zeta) \in \zeta$ for μ-a.a. $\zeta \in \widehat{R}$ and $\rho'(\zeta) \in \zeta$ for ν-a.a. $\zeta \in \widehat{R}$. From the form of the coherence

28. MEASURES ON THE QUASI-DUAL AND REPRESENTATIONS

appearing in the direct integral representations $\tau(\mu,\pi)$, $\tau(\mu,\pi')$, $\tau(\nu,\rho)$, and $\tau(\nu,\rho')$ it is clear that $\tau(\mu,\pi')$ and $\tau(\nu,\rho')$ are central decompositions of $\tau(\mu,\pi)(\cdot) \otimes I$ and $\tau(\nu,\rho)(\cdot) \otimes I$, respectively (cf. Lemma 7.4). Now $\tau(\mu,\pi)(\cdot) \otimes I$ and $\tau(\nu,\rho)(\cdot) \otimes I$ are equivalent by Proposition A.3, and hence μ and ν are equivalent by Theorem 28.4.

The proof of the remaining two assertions are similar to those of the corresponding parts of Theorem 27.2 and will be omitted. □

The promised example of a noncanonical standard Borel measure on a quasi-dual will now be described. Let G be the discrete group constructed in Example 10.7, let λ be the left regular representation of G, and let $R = \ell^1(G)$; R is then a separable involutive Banach algebra with an identity. It is well-known from the theory of von Neumann algebras that $\lambda(G)''$ is a factor of type II_1, and, moreover, one knows from Proposition 11.2, Example 15.2, and Theorem C.2 that there is a standard Borel space Z, a finite Borel measure ν on Z, a Borel field of Hilbert spaces H on Z, a coherence α for H, and an α-Borel field of representations ρ of R over H such that the $\rho(\zeta)$, $\zeta \in Z$, are irreducible and mutually disjoint and such that $\int_Z^\alpha \rho(\zeta)d\nu(\zeta)$ is a direct integral decomposition of $\tilde{\lambda}$ with respect to an abelian von Neumann subalgebra M of $\tilde{\lambda}(R)'$. Imitating the proof of Theorem 28.4, one can find a standard Borel measure μ on $\widehat{R}$ and a Borel function π from $\widehat{R}$ to Irr R with $\pi(\zeta) \in \zeta$ for μ-a.a. $\zeta \in \widehat{R}$ and such that $\tau(\mu,\pi)$ is a direct integral decomposition of $\tilde{\lambda}$ with respect to M. Thus $\tau(\mu,\pi)$ is a maximal and not a central decomposition of $\tilde{\lambda}$, and hence μ is not canonical.

Finally, suppose that one is given a separable involutive Banach algebra R of type I and a representation σ of R on a separable Hilbert space. Let μ be some member of the Borel measure class on $\widehat{R}$ determined by σ in accordance with Theorem 28.4. Next, choose a multiplicity-free representation σ' of R which is quasi-equivalent to σ and acts on a separable Hilbert space, and let μ' be a member of the Borel measure class on $\widehat{R}$ determined by σ' in accordance with Theorem 27.1. A natural question to consider is that of how μ and μ' are related. In answering this question it turns out that one may as well consider the more general situation in which only the representation σ (and not the algebra R) is assumed to be of type I.

Accordingly, let R be a separable involutive Banach algebra, let p and q be the canonical mappings of Irr R onto $\widehat{R}$ and of Fac R onto $\widecheck{R}$, respectively, and let θ be the inclusion mapping of Irr R into Fac R. Then as two irreducible representations are equivalent if and only if they are quasi-equivalent (this follows from Proposition A.3), there is a one-one Borel mapping $\tilde{\theta}$ of $\hat{R}$ into $\check{R}$ such that $\tilde{\theta} \circ p = q \circ \theta$. Now consider a type I representation σ of R on a separable Hilbert space. Then according to Theorem 28.4 there is a finite Borel measure μ on $\check{R}$ which is standard and canonical and there is a Borel map ρ from $\check{R}$ to Fac R such that $\rho(\zeta) \in \zeta$ for μ-a.a. $\zeta \in \check{R}$ and such that $\tau(\mu,\rho)$ is a central decomposition of σ. Next, by Theorem A.6 there is a multiplicity-free representation σ' of R on a separable Hilbert space which is quasi-equivalent to σ. Then by Theorems 12.1, 13.1, and 13.3 there is a standard Borel space Z, a finite Borel measure ν on Z, a Borel map π from Z to Irr R such that the representations $\pi(\zeta)$, $\zeta \in Z$, are mutually disjoint and such that if $\alpha(\zeta)$ is the identity operator on the Hilbert space on which $\pi(\zeta)$ acts for each $\zeta \in Z$ then $\int_Z^\alpha \pi(\zeta)d\nu(\zeta)$ is a central decomposition of σ'. Notice that if S is a Borel subset of Z then $\theta(\pi(S))$ is, by Theorem 2.13, a Borel subset of Fac R meeting each quasi-equivalence class in Fac R in at most one point, and hence

$$p^{-1}(p(\pi(S))) = \{\tau \in \text{Irr } R: \tau \simeq \tau' \text{ for some } \tau' \in \pi(S)\}$$
$$= \theta^{-1}(\{\tau \in \text{Fac } R: \tau \approx \tau' \text{ for some } \tau' \in \theta(\pi(S))\})$$

is a Borel subset of Irr R by Proposition 28.3. This means that $p(\pi(Z))$ is a Borel subset of $\hat{R}$ and that $p \circ \pi$ is a Borel isomorphism of Z onto $p(\pi(Z))$ with respect to the relative Borel structure on $p(\pi(Z))$. Put $\mu' = (p \circ \pi)_*(\nu)$ and let π' be any Borel map from $\hat{R}$ to Irr R which agrees with $\pi \circ (p \circ \pi)^{-1}$ on $p(\pi(Z))$. Then μ' is a standard Borel measure on $\hat{R}$, $\pi'(\zeta) \in \zeta$ for μ'-a.a. $\zeta \in \hat{R}$, and $\int_{\hat{R}}^\beta \pi'(\zeta) d\mu'(\zeta)$ is a central decomposition of σ'; here $\beta(\zeta)$ is, for each $\zeta \in \hat{R}$, the identity operator on the Hilbert space on which $\pi'(\zeta)$ acts. The next step is to show that the measures $\tilde{\theta}_*(\mu')$ and μ are necessarily equivalent; this will essentially answer the question posed in the preceeding paragraph.

It follows easily from Proposition 19.3 and Theorem 21.2 that the representation $\rho(\zeta)$ is of type I for μ-a.a. $\zeta \in \check{R}$. Choose a standard Borel subset W of $\check{R}$ whose complement is μ-null and which is such that $\rho(\zeta) \in \zeta$ and $\rho(\zeta)$ is of type I for each $\zeta \in W$. Then $\theta^{-1}(q^{-1}(W))$ is a Borel subset of Irr R whose image under $q \circ \theta$ is all of W, and hence there is (by Theorem 4.3) a Borel map ρ' from $\check{R}$ to Irr R with $(q \circ \theta \circ \rho')(\zeta) = \zeta$ for μ-a.a. $\zeta \in \check{R}$. Letting $\gamma(\zeta)$ be the identity operator on the Hilbert space on which $\rho'(\zeta)$ acts for each $\zeta \in \check{R}$, the two representations $\tau(\mu,\rho)$ and $\int_{\check{R}}^\gamma \rho'(\zeta)d\mu(\zeta)$ are quasi-equivalent by Theorem 12.1, and, moreover, one can mimic the first

28. MEASURES ON THE QUASI-DUAL AND REPRESENTATIONS 135

part of the proof of Theorem 27.1 to deduce that $\int_{\hat{R}}^{\gamma} \rho'(\zeta)d\mu(\zeta)$ is multiplicity-free. But then σ' and $\int_{\hat{R}}^{\gamma} \rho'(\zeta)d\mu(\zeta)$ must be equivalent by Proposition A.3, and therefore $\int_{\hat{R}}^{\gamma} \rho'(\zeta)d\mu(\zeta)$ is a central decomposition of σ'. As the measures μ and μ' are both standard it now follows from Theorems 9.2 and 12.4 that there is a Borel map ω from $\check{R}$ to $\hat{R}$ with the following properties:

(1) $\omega_*(\mu')$ is equivalent to μ,
(2) there are standard Borel subsets S and T of $\check{R}$ and $\hat{R}$ whose complements are μ'-null and μ-null, respectively, and such that the restriction of ω to S is a Borel isomorphism of S onto T,
(3) $\pi'(\zeta) \simeq \rho'(\omega(\zeta))$ for each $\zeta \in S$, and
(4) $\pi'(\zeta) \in \zeta$ for all $\zeta \in S$ and $\rho(\zeta) \in \zeta$ for all $\zeta \in T$.

But then for any point ζ in S one has

$$\omega(\zeta) = q(\rho(\omega(\zeta)))$$
$$= q(\theta(\rho'(\omega(\zeta))))$$
$$= \tilde{\theta}(p(\pi'(\zeta)))$$
$$= \tilde{\theta}(\zeta)$$

and hence $\tilde{\theta}_*(\mu')$ and μ are equivalent. From this and the fact that $p \circ \rho' \circ \tilde{\theta}$ is the identity on S it follows easily that the measures $(p \circ \rho')_*(\mu)$ and μ' are equivalent.

If the algebra R were of type I (or equivalently, had a smooth dual) then μ' and μ belong to the Borel measure classes on $\check{R}$ and $\hat{R}$ associated with σ' and σ in accordance with Theorems 27.1 and 28.4, respectively. This discussion, therefore, gives a complete answer to the question as to how these two classes are related.

HISTORICAL COMMENTS

The results in this chapter which relate the Mackey Borel structures and the Borel measures on the dual and the quasi-dual of a separable involutive Banach algebra to the representation theory of that algebra are largely due to Mackey and Ernest and have been included in the books by Arverson [1], Dixmier [9], and Pedersen [52]. The reader who wishes to learn more about this subject would be well-advised to consult the last-mentioned of these books.

What is now called the Mackey Borel structure on $\check{R}$ was introduced and first studied by Mackey in [41, Section 8]. Here he showed, in particular, that $\text{Rep}_n R$ for all $n \in \mathbb{N} \cup \{\infty\}$ and Irr R were standard Borel spaces and that one-point subsets of $\hat{R}$ are Borel. (Dixmier subsequently showed that $\text{Rep}_n R$ is actually a Polish space in the given topology [7, Lemme 1]). Later in this same paper Mackey went on to derive most of what is in Section 27 [41, Section 10].

The Mackey Borel structure on $\hat{R}$ was introduced and first studied by Ernest in [18, Section 2]. He showed, among other things, that Fac R is a standard Borel space and that one-point subsets of $\hat{R}$ are Borel, and he proved Lemma 28.3. Later in this paper Ernest obtained Theorems 28.4 and 28.5 [18, Sections 4,5].

Lemmas 26.5 and 28.2 are due to Mackey [39, Theorems 2.7 and 2.8] and Proposition 26.6 and Lemma 28.1 to Effros [15, Lemma 4.2; 12, Theorem 4.1].

The statement that a separable involutive Banach algebra with an approximate identity has a smooth dual if and only if it is of type I was conjectured by Mackey [40, Section 2.3; 41, p. 163] and proven by Glimm [20, Theorem 2]. Effros later gave a simpler proof of this important result [15, Theorem 4.3].

Canonical measures were introduced by Ernest [18, Section 5] by means of a definition slightly different from the one employed here. The example of a noncanonical measure given in Section 28 is due to Ernest [18, pp. 269-270]. The problem of characterizing the canonical measures among all of the standard Borel measures on $\hat{R}$ in terms of the Mackey Borel structure on $\hat{R}$ was solved by Effros [17].

The final assertion of Theorem 26.1 is due to Dixmier, Feldman, and Nielsen. More precisely, if

$$W_k = \{\pi \in \text{Rep } R: \pi(R)'' \text{ is } P_k\}, \quad 1 \leq k \leq 9$$

then Dixmier showed that W_9 and $W_9 \cap W_1$ are Borel sets [18, Theorem 1; 8, Corollaire 1], Guichardet showed that $W_9 \cap W_2 \cap W_5$ is a Borel set [23, Theorem 1 of Section I.2], Feldman showed that $W_9 \cap W_5$ is a Borel set, that $W_9 \cap W_2$ is an analytic set, and that $W_9 \cap W_3$ is a coanalytic set [19, Lemmas 4 and 5], and Nielsen showed that each of the sets W_k is actually Borel [51, Section 4]. To each state ϕ on R one can associate a representation π_ϕ of R by means of the GNS-construction, and as each W_k is Borel one can deduce that the set Σ_k of states ϕ on R for which $\pi_\phi(R)''$ is P_k is Borel for $1 \leq k \leq 9$ (cf. [19] or [51, Section 4]). On the other hand, Pedersen [52, Theorem 5.7.4] has found a proof of the fact that the sets Σ_9, $\Sigma_9 \cap \Sigma_1$, $\Sigma_9 \cap \Sigma_2$, $\Sigma_9 \cap \Sigma_2 \cap \Sigma_5$, and $\Sigma_9 \cap \Sigma_3$ are all Borel which avoids both Theorem 21.1 and

28. MEASURES ON THE QUASI-DUAL AND REPRESENTATIONS

modular automorphism groups. If one now takes R to be the convolution algebra of the free group on infinitely many generators one could apply this result of Pedersen and Lemma 24.4 to deduce, first, Lemma 24.7, then Theorem 21.1(i), and, finally, that each of the sets W_k is Borel.

Appendix A
VON NEUMANN ALGEBRAS

The reader is assumed to be somewhat familiar with the theory of von Neumann algebras and this appendix merely contains a list of those definitions and results which are needed in discussing direct integrals. The standard reference for the theory of von Neumann algebras is, of course, Diximer's book [10], and the proofs of all of the results mentioned here with the exception of those concerning modular automorphism groups can be found therein; for the theory of modular automorphism groups the reader should consult Takesaki's notes [56]. It should be pointed out that the definitions given here in connection with the classification theory of von Neumann algebras are different from but equivalent to those in Dixmier's book in that it is the comparison of projections rather than the existence of traces which is taken to be fundamental.

A *von Neumann algebra* on a Hilbert space H is by definition a strongly closed *-algebra of operators of H containing the identity operator. Recall that the commutant S' of a nonempty set S of operators on H consists of all those operators on H commuting with each element of S; it is clear that S' is a von Neumann algebra if S is self-adjoint and that $S \cap S'$ is the center of S if S is an algebra. The first of the theorems quoted below implies that a *-algebra of operators A on H containing the identity operator is a von Neumann algebra if and only if it satisfies $A = A''$ and that if S is a self-adjoint set of operators on H then S'' is the smallest von Neumann algebra on H containing S.

THEOREM A.1. (von Neumann density theorem). If A is a *-algebra of operators on a Hilbert space then the operators of the form $A + aI$, $A \in A$ and $a \in C$, are strongly dense in A''.

THEOREM A.2. (Kaplansky density theorem). Suppose that A is a *-algebra of operators on a separable Hilbert space. Then given an operator S in A'', there are sequences (A_n) in A and (a_n) in $\mathbb{C}$ such that $\|A_n + a_n I\| \leq \|S\|$ for all n and $A_n + a_n I \to S$ strongly as $n \to \infty$.

Given von Neumann algebras A and B acting on Hilbert spaces H and K, respectively, a *-isomorphism θ of A onto B is said to be *spatial* if there is a linear isometry U of H onto K with $\theta(A) = UAU^*$ for all $A \in A$.

PROPOSITION A.3. If θ is a *-isomorphism of a von Neumann algebra A onto a von Neumann algebra B then:

(i) θ is isometric,
(ii) θ is spatial if both A' and B' are abelian, and
(iii) there is a Hilbert space H (which can be taken to be separable if both A and B act on separable Hilbert spaces) such that the *-isomorphism $A \otimes I \mapsto \theta(A) \otimes I$ of $A \otimes C(H)$ onto $B \otimes C(H)$ is spatial.

PROPOSITION A.4. An abelian von Neumann algebra acting on a separable Hilbert space is *-isomorphic to $L^\infty(\mu)$ for some Borel measure μ on the closed unit interval.

Actually, the last assertion of Proposition A.3 does not seem to appear explicitly in Dixmier's book [10], but can easily be deduced from the structure theorem for *-isomorphisms contained therein. This reduction, however, requires the notion of equivalence of projections and so will be deferred for the moment.

Consider a von Neumann algebra A acting on a Hilbert space H and a projection E in A. For any operator A on H let A_E denote the restriction of EA to the range of E, and put $S_E = \{A_E : A \in S\}$ for any subset S of $L(H)$; then $A_E \in L(EH)$ and $S_E \subset L(EH)$. The smallest projection on H majorizing UEU^* for each unitary operator U in A is equal to the smallest projection in $A \cap A'$ majorizing E and is called the *central support* of E (in A). A second projection F in A is said to be *equivalent* to E (in A) if there is an operator V in A with $V^*V = E$ and $VV^* = F$, and in this case one writes $E \sim F$; notice that such an operator V is necessarily a partial isometry. The projection E is called *abelian* if A_E is abelian, *infinite* if there is a projection F in A with $E \sim F < E$, and *finite* if it is not infinite. The algebra A itself is said to be *continuous* if each projection in A is the sum of two equivalent projections in A, *finite* or *infinite* according to whether I is finite or infinite, respectively, *semi-finite* if it contains a finite projection whose central support is I, and *properly infinite* [respectively, *purely infinite*] if each nonzero central projection [respectively, each nonzero projection] in A is infinite. In addition, A is said to be *of type* I if it contains an abelian projection whose central support is I, *of type* II if it is continuous and semi-finite, *of type* II_1 [respectively, II_∞] if it is of type II and is finite [respectively, infinite], and *of type* III if it is purely infinite. Finally, A is said to be of type I_a, where a is some nonzero cardinal number, if there is a set J of cardinality a and a family $(E_j)_{j \in J}$ of mutually equivalent and mutually orthogonal abelian projections in A with $I = \sum_{j \in J} E_j$, or equivalently, if A is *-isomorphic to $M \otimes L(H_a)$ for some abelian von Neumann algebra M and some Hilbert space H_a of dimension a.

The proof of Proposition A.3 (iii) which was just alluded to can now be described. If A, B, and θ are as in the proposition then the structure theorem for *-isomorphisms implies that there is a Hilbert space H (which can be taken to be separable if both A and B act on separable Hilbert spaces) and projections E and F in $[A \otimes C(H)]'$ and $[B \otimes C(H)]'$, respectively, such that the maps $A \otimes I \mapsto (\theta(A) \otimes I)_F$ and $B \otimes I \mapsto (\theta^{-1}(B) \otimes I)_E$ are spatial isomorphisms of $A \otimes C(H)$ onto $[B \otimes C(H)]_F$ and of $B \otimes C(H)$ onto $[A \otimes C(H)]_E$, respectively. Then one can show that the four projections $I \oplus 0$, $E \oplus 0$, $0 \oplus I$, and $0 \oplus F$ belong to the von Neumann algebra $\{(A \otimes I) \oplus (\theta(A) \otimes I) : A \in A\}'$ and that in this algebra $I \oplus 0$ is equivalent to $0 \oplus F$ and $0 \oplus I$ is equivalent to $E \oplus 0$. But then $I \oplus 0$ and $0 \oplus I$ will be equivalent, and from this it is easy to deduce that $A \otimes I \mapsto \theta(A) \otimes I$ is a spatial isomorphism of $A \otimes C(H)$ onto $B \otimes C(H)$.

PROPOSITION A.5. Let A be a von Neumann algebra and let E be a projection belonging to either A or A'. Then both A_E and $(A')_E$ are von Neumann algebras, $(A_E)' = (A')_E$, and $(M_E)'' = A_E$ if M is a self-adjoint subset of A with $M'' = A$. Moreover, if E belongs to A' and if its central support in A' is F then the map $A_F \mapsto A_E$ is a *-isomorphism of A_F onto A_E.

By definition, the *reduction* of a von Neumann algebra A by a nonzero projection E in A is the von Neumann algebra A_E.

THEOREM A.6. For a von Neumann algebra A acting on a separable Hilbert space the following are equivalent:

(a) A is of type I,

(b) A is *-isomorphic to a von Neumann algebra acting on a separable Hilbert space and having an abelian commutant, and

(c) there is a subset J of $\mathbb{N} \cup \{\infty\}$ and for each n in J a von Neumann algebra A_n of type I_n such that A and $\oplus_{n \in J} A_n$ are spatially isomorphic.

PROPOSITION A.7. For a von Neumann algebra A acting on a separable Hilbert space the following are equivalent:

(a) A is semi-finite, and

(b) A is *-isomorphic to a von Neumann algebra acting on a separable Hilbert space and having a finite commutant.

THEOREM A.8. A von Neumann algebra A is continuous if and only if A_E fails to be of type I for every nonzero central projection E in A.

THEOREM A.9. If a von Neumann algebra is type I, type II, type III, continuous, or semi-finite then so is its commutant.

THEOREM A.10. For any von Neumann algebra A there are uniquely-determined central projections E, F and G with the properties that for any nonzero central projection H in A, A_H is of type I if $H \leq E$, continuous if $H \leq I - E$, finite if $H \leq F$, properly infinite if $H \leq I - F$, semi-finite if $H \leq G$, and purely infinite if $H \leq I - G$.

Consider again a von Neumann algebra A acting on a Hilbert space H. A linear functional ϕ on A is said to be *positive* if it maps the set A^+ of all nonnegative operators in A into $[0,\infty)$, and in this case it is called a *state* if $\phi(I) = 1$, *faithful* if $A \in A^+$ and $\phi(A) = 0$ implies $A = 0$, and a *trace* if $\phi(AB) = \phi(BA)$ for all $A, B \in A$. The set $L(H)_*$ of all ultraweakly continuous linear functionals on $L(H)$ is a norm-closed subspace of the dual space of $L(H)$ and so is a complex Banach space and, moreover, the canonical bilinear form on $L(H)_* \times L(H)$ induces an isometric isomorphism between $L(H)$ and the dual space of $L(H)_*$ whose associated weak *-topology on $L(H)$ is just the ultraweak topology. The set of all positive elements of $L(H)_*$ will be denoted by $L(H)_*^+$.

THEOREM A.11. A von Neumann algebra A acting on a separable Hilbert space H is finite if and only if there is an element of $L(H)_*^+$ whose restriction to A is a faithful trace.

A vector x in a Hilbert space H on which a von Neumann algebra A acts is said to be *cyclic* for A if the set Ax is dense in H and *separating* for A if $A \in A$ and $Ax = 0$ implies $A = 0$. It is not hard to prove that a vector is cyclic for A if and only if it is separating for A'. An *automorphism* of A is by definition a *-isomorphism of A onto itself. An automorphism θ of A is said to be *inner* if there is a unitary operator U in A with $UAU^* = \theta(A)$ for all $A \in A$, and *outer* if it is not inner. A *one-parameter group of automorphisms* of A is a family (σ_t), or more precisely, a family $(\sigma_t)_{t \in \mathbb{R}}$, of automorphisms of A such that the map $t \mapsto \sigma_t$ is a homomorphism from $\mathbb{R}$ to the group of automorphisms of A; such a family (σ_t) is called *strongly continuous* if the map $t \mapsto \sigma_t(A)$ from $\mathbb{R}$ to A is continuous with respect to the strong operator topology on A for each fixed $A \in A$.

Modular automorphism groups are used only once, in the proof of Lemma 24.7, and the following summary is very incomplete. Consider once again a von Neumann algebra A acting on a separable Hilbert space H, and suppose that x is a vector in H which is cyclic and separating for A. Let ϕ denote the state $A \mapsto \langle Ax, x \rangle$ on A. Then there is a unique strongly continuous one-parameter group of automorphisms (σ_t) of A satisfying the following so-called KMS-*condition*: given any two operators A and B in A, there is a continuous bounded complex-valued function f on the strip $S = \{z \in \mathbb{C}: 0 \leq \text{Im}(z) \leq 1\}$ which is analytic on the interior of S and satisfies the boundary conditions

$$f(t) = \phi(A\sigma_t(B)) \quad \text{and} \quad f(t+i) = \phi(\sigma_t(B)A)$$

for all real t. This group (σ_t) is called the *modular automorphism group of* A *associated with* ϕ. Taking $A = I$ in the KMS-condition leads to the fact that $\phi \circ \sigma_t = \phi$ for all real t (use Liouville's theorem). The importance of the modular automorphism group is due in part to the following facts:

(i) if A is not semi-finite then σ_s is outer for some $s \in \mathbb{R}$, and

(ii) if A is semi-finite then there is a strongly continuous one-parameter unitary group (U_t) in A with $\sigma_t(A) = U_t A U_t^*$ for all $A \in A$ and $t \in \mathbb{R}$.

Appendix B

REPRESENTATIONS OF INVOLUTIVE BANACH ALGEBRAS

The purpose of this appendix is to give a self-contained account of that very small initial segment of the representation theory of involutive Banach algebras which is needed in reading these notes. Part or all of the material discussed here appears in a number of books, including those by Arveson [1], Dixmier [9], Naimark [47], and Pedersen [52].

An *involutive Banach algebra* is by definition a complex Banach algebra R on which there is defined a mapping $R \mapsto R*$ (called the *involution* of R) which is anti-linear, isometric, and satisfies $R** = R$ and $(RS)* = S*R*$ for all $R,S \in R$. The two most important examples of such algebras are the convolution algebras of locally compact groups (these are discussed in the next appendix) and the C*-algebras. A *C*-algebra* is an involutive Banach algebra in which the involution and norm are related by the identity $\|R*R\| = \|R\|^2$, $R \in R$. Any norm-closed *-algebra of operators on a Hilbert space is a C*-algebra, and there is an important theorem (which will not be needed in these notes) to the effect that every C*-algebra is isometrically *-isomorphic to one of this form.

A *representation* of an involutive Banach algebra R on a Hilbert space H is by definition a *-homomorphism of R into $L(H)$. The following result is basic and is used repeatedly in Chapter 3.

PROPOSITION B.1. Any representation of an involutive Banach algebra is norm-decreasing.

Proof. Let π be a representation of an involutive Banach algebra R on a Hilbert space H, and let R be some element of R. If a is a real number with $a > \|R*R\|$ then the series $\sum_{n=1}^{\infty} a^{-n}(R*R)^n$ converges in norm to an element,

145

say S, of R which must satisfy the equation

$$aS - R^*R = R^*RS = SR^*R$$

Using this equation one can show that $-\frac{1}{a}[\pi(S) + I]$ is an inverse of $\pi(R^*R) - aI$. Thus the resolvent set of $\pi(R^*R) = \pi(R)^*\pi(R)$ contains the interval $(\|R^*R\|, \infty)$, hence $\|\pi(R^*R)\| \leq \|R^*R\|$, and therefore

$$\|\pi(R)\|^2 = \|\pi(R^*R)\| \leq \|R^*R\| \leq \|R\|^2 \quad \square$$

Consider an involutive algebra R. Give a nonempty set J and, for each j in J, a representation π_j of R on a Hilbert space H_j, one can (by Proposition B.1) form the representation $R \mapsto \oplus_{j \in J} \pi_j(R)$ of R on $\oplus_{j \in J} H_j$; this representation is called the *direct sum* of the π_j, $j \in J$, and is denoted by $\oplus_{j \in J} \pi_j$. If $J = \{1,\ldots,n\}$ or if $J = \mathbb{N}$ then of course one writes $\oplus_{j=1}^{n} \pi_j$ or $\oplus_{j=1}^{\infty} \pi_j$, respectively, instead of $\oplus_{j \in J} \pi_j$. The *zero-representation* of R on a Hilbert space H is the function which carries every element of R onto the zero-operator on H. Now consider a representation π of R on a Hilbert space H. If E is a projection in $\pi(R)'$ then each $\pi(R)$, $R \in R$, leaves the range of E invariant and hence $R \mapsto \pi(R)|_{EH}$ is a representation of R on EH; this representation is called the *subrepresentation* of π determined by E and is denoted by π_E. One says that π is *nondegenerate* if the vectors $\pi(R)x$, $R \in R$ and $x \in H$, are total in H, *irreducible* if $\pi(R)'' = L(H)$, *multiplicity-free* if $\pi(R)'$ is abelian, a *factor representation* or *factorial* if $\pi(R)''$ is a factor, and of *type* x if $\pi(R)''$ is of type x, x = I, II_1, II_∞, or III. Notice that π is nondegenerate if and only if $\cap_{R \in R} \ker \pi(R) = \{0\}$ and irreducible if and only if the only closed subspaces of H invariant under the $\pi(R)$, $R \in R$, are $\{0\}$ and H. In particular, if π is irreducible and if $\dim H \geq 2$ then π is nondegenerate, while π is always irreducible if H is 0- or 1-dimensional. Finally, consider a second representation ρ of R on a Hilbert space K. An operator A from H to K is said to *intertwine* π and ρ if $A\pi(R) = \rho(R)A$ for all $R \in R$. One calls π and ρ *quasi-equivalent* if there is a *-isomorphism θ of $\pi(R)''$ onto $\rho(R)''$ satisfying $\theta \circ \pi = \rho$, *equivalent* if there is a linear isometry of H onto K intertwining π and ρ, and *disjoint* if there is no nonzero operator from H to K intertwining π and ρ. It is customary to write $\pi \approx \rho$ if π and ρ are quasi-equivalent, $\pi \simeq \rho$ if they are equivalent, $\pi \, \natural \, \rho$ if they are disjoint, and $\pi \leq \rho$ if π is equivalent to a subrepresentation of ρ. Notice that π and ρ are quasi-equivalent if they are equivalent, that $\pi(R)'$ consists of the operators on H intertwining π and π, and that an operator A from H to K intertwines π and ρ if and only if A^* intertwines ρ and π. The best way to illustrate the difference between equivalence and quasi-equivalence is to remark that if π is a nonzero irreducible representation of R then π and $\pi \oplus \pi$ are quasi-equivalent but not equivalent and that if π and ρ are representations of R on separable Hilbert spaces then π and ρ are quasi-equivalent if and only if the representations $R \mapsto \pi(R) \otimes I$ and $R \mapsto \rho(R) \otimes I$ are equivalent, where I is the identity operator on some separable infinite-dimensional Hilbert space (this follows from Proposition A.3).

PROPOSITION B.2. Every representation of an involutive Banach algebra is quasi-equivalent to a nondegenerate representation and is equivalent to the direct sum of a nondegenerate representation and a zero-representation.

Proof. Let π be a representation of an involutive Banach algebra R on a Hilbert space H, and let E be the projection of H onto the closed subspace spanned by the vectors $\pi(R)x$, $R \in R$ and $x \in H$. Then $\pi(R)EH \subset EH$, i.e., $E\pi(R)E = \pi(R)E$, for all $R \in R$, and hence $E \in \pi(R)'$. One can therefore form the subrepresentations π_E and π_{I-E}, and obviously $\pi \approx \pi_E$, $\pi \simeq \pi_E \oplus \pi_{I-E}$, π_E is nondegenerate, and π_{I-E} is a zero-representation. □

PROPOSITION B.3. For representations π and ρ of an involutive Banach algebra R on Hilbert spaces H and K, respectively, the following conditions are equivalent:

(i) π and ρ fail to be disjoint,

(ii) there is a nonzero partially isometric operator from H to K intertwining π and ρ, and

(iii) there are nonzero projections E and F in $\pi(R)'$ and $\rho(R)'$, respectively, such that π_E and ρ_F are equivalent.

Proof. (i) $\Rightarrow$ (ii) If π and ρ fail to be disjoint there is, by definition, a nonzero operator A from H to K intertwining π and ρ. It is easy to see that $\|Ax\| = \|(A^*)^{\frac{1}{2}}x\|$ for all $x \in H$, and hence that there is a partially isometric operator V from H to K satisfying $A = V(A^*A)^{\frac{1}{2}}$ and $V(I-E) = 0$, where E is the projection of H onto the closure of the range of $(A^*A)^{\frac{1}{2}}$. Then V is clearly nonzero, and E lies in $\pi(R)'$ since A^*A does. Now given a vector x in H, one can find a sequence (x_n) in H with $Ex = \lim_{n\to\infty}(A^*A)^{\frac{1}{2}}x_n$, and thus

$$V\pi(R)x = VE\pi(R)x = \lim_{n\to\infty}V\pi(R)(A^*A)^{\frac{1}{2}}x_n = \lim_{n\to\infty}V(A^*A)^{\frac{1}{2}}\pi(R)x_n =$$
$$\lim_{n\to\infty}\rho(R)V(A^*A)^{\frac{1}{2}}x_n = \rho(R)Vx \quad \text{for all } R \in R.$$

(ii) $\Rightarrow$ (iii) Suppose that V is a nonzero partially isometric operator from H to K intertwining π and ρ. Letting E and F be the projections of H and K onto V^*H and VK, respectively, one knows that $0 \neq V^*V = E \in \pi(R)'$ and that $0 \neq VV^* = F \in \rho(R)'$. Moreover, it is clear that the restriction of V to EH is a linear isometry of EH onto FK intertwining π_E and ρ_F, and hence that $\pi_E \simeq \rho_F$.

(iii) $\Rightarrow$ (i) Let E and F be as in (iii), and let V be a linear isometry of EH onto FK intertwining π_E and ρ_F. Then VE is a nonzero operator from H to K, and an easy calculation will show that VE intertwines π and ρ. □

COROLLARY B.4. (Schur's Lemma). Two irreducible representations of an involutive Banach algebra are either equivalent or else disjoint.

Proof. This is an obvious consequence of the Proposition. □

PROPOSITION B.5. Two representations of an involutive Banach algebra cannot be both quasi-equivalent and disjoint, and two factor representations must be either quasi-equivalent or else disjoint.

Proof. Let π and ρ be two representations of an involutive Banach algebra, and suppose that π and ρ are quasi-equivalent. Then if I denotes the identity operator on a suitable Hilbert space K the representations $\rho(\cdot) \otimes I$ and $\rho(\cdot) \otimes I$ are equivalent by Proposition A.3. Let H_π and H_ρ denote the Hilbert spaces on which π and ρ act, respectively, and let W be a linear isometry of $H_\pi \otimes K$ onto $H_\rho \otimes K$ which intertwines $\pi(\cdot) \otimes I$ and $\rho(\cdot) \otimes I$. If z_1 is some unit vector in K then there is a unit vector z_2 in K such that if P denotes the isometry $y \mapsto y \otimes z_2$ of H_ρ into $H_\rho \otimes K$ then the map $x \mapsto P^*W(x \otimes z_1)$ from H_π to H_ρ is nonzero. But this map is easily seen to intertwine π and ρ, and therefore π and ρ are not disjoint.

Turning now to the second assertion, suppose that π and ρ are factorial and not disjoint. Then by Proposition B.3 there must be nonzero projections E and F in $\pi(R)'$ and $\rho(R)'$, respectively, such that $\pi_E \simeq \pi_F$. Now the central support of E in $\pi(R)'$ is I as π is factorial, hence $A \mapsto A_E$ is a *-isomorphism of $\pi(R)''$ onto $\pi(R)''_E = \pi_E(R)''$ by Proposition A.5, and therefore $\pi \approx \pi_E$. As a similar argument will show that $\rho \approx \rho_F$, it is clear that π and ρ must be quasi-equivalent. □

An *approximate identity* for an involutive Banach algebra R is by definition a net $(R_a)_{a \in A}$ of self-adjoint elements in the unit ball of R satisfying

REPRESENTATIONS OF INVOLUTIVE BANACH ALGEBRAS

$$\lim_a R_a R = \lim_a RR_a = R \quad \text{for all } R \in \mathcal{R}$$

It will be convenient to call an approximate identity *sequential* if it is a sequence.

PROPOSITION B.6. Let $\mathcal{R}$ be an involutive Banach algebra, let $(R_a)_{a \in A}$ be an approximate identity for $\mathcal{R}$, and let π be a representation of $\mathcal{R}$ on a Hilbert space H. Then the net $(\pi(R_a))_{a \in A}$ converges strongly to the projection of H onto the closed subspace spanned by the vectors $\pi(R)x$, $R \in \mathcal{R}$ and $x \in H$.

Proof. Let E be the projection in question and consider a vector x in H. Given an $\varepsilon > 0$, there will be elements $R_1, \ldots, R_n$ in $\mathcal{R}$ and vectors $y_1, \ldots, y_n$ in H such that $\| Ex - \sum_{j=1}^n \pi(R_j) y_j \| < \varepsilon$. Now using Proposition B.1 and the fact that $\pi(R)E = \pi(R)$, $R \in \mathcal{R}$, (cf. the proof of Proposition B.2) one obtains

$$\| Ex - \pi(R_a)x \| \leq \| Ex - \sum_{j=1}^n \pi(R_j) y_j \| + \| \sum_{j=1}^n \pi(R_j - R_a R_j) y_j \|$$

$$+ \| \pi(R_a) \sum_{j=1}^n \pi(R_j) y_j - \pi(R_a) Ex \|$$

$$< 2\varepsilon + \sum_{j=1}^n \| R_j - R_a R_j \| \, \| y_j \|$$

for all $a \in A$, which is enough to prove the lemma. $\square$

Appendix C
REPRESENTATIONS OF LOCALLY COMPACT GROUPS

This appendix contains as much of the representation theory of locally compact groups as is needed to read these notes. Except for the actual definition of a representation of a locally compact group, this material is all standard and can be found in the books by Dixmier [9], Loomis [35], and Naimark [47].

Suppose that G is a locally compact group. Let Δ be the modular function of G, let $\int_G f(r)dr$ or simply $\int f(r)dr$ denote the integral (assuming it exists) of a function f on G with respect to some choice of a left-invariant Haar measure on G, and let $L^p(G)$, $1 \leq p \leq \infty$, be the usual L^p-spaces associated with this measure. In this appendix the L^p- and L^p-spaces with respect to the fixed Haar measure on G will be identified, and thus elements of $L^p(G)$ will be regarded as functions on G; this is contrary to the convention made in Section 4 but will be convenient and should not lead to any confusion. The two formulae

$$(f*g)(r) = \int f(s)g(s^{-1}r)ds$$
$$f*(r) = \Delta(r^{-1})\bar{f}(r^{-1})$$

where r in G and f,g in $L^1(G)$ are arbitrary, define a multiplication (called *convolution*) and an involution, respectively, on $L^1(G)$ making $L^1(G)$ into an involutive Banach algebra. The only difficulty in verifying this assertion is in showing that the integral defining $(f*g)(r)$ exists for almost all r, that f*g lies in $L^1(G)$, and that $\|f*g\| \leq \|f\| \|g\|$. These three points can all be resolved by noting that if f and g are in $L^1(G)$ then

$$\iint |f(s)g(s^{-1}r)|drds = \iint |f(s)g(r)|drds$$
$$= \|f\| \, \|g\|$$

and hence (by Fubini's theorem, which is applicable here since f and g, being in $L^1(G)$, are zero off σ-finite subsets of G) that $s \mapsto f(s)g(s^{-1}r)$ is in $L^1(G)$ for almost all r and that $\|f*g\| \leq \|f\| \, \|g\|$. It is clear that $L^1(G)$ is separable [respectively, is abelian, has an identity] if G is separable [respectively, abelian, discrete]. Finally, $L^1(G)$ is in general not a C*-algebra (to see this one need only look at the case where G is the two-element group).

PROPOSITION C.1. $L^1(G)$ has an approximate identity which may be taken to be a sequence if G is separable.

Proof. Let $\mathcal{U}$ be a base for the neighbourhoods of e, the identity in G, consisting of sets which are open, symmetric, and relatively compact. Writing $U \leq V$ if U and V are sets in $\mathcal{U}$ with $U \supset V$, $(\mathcal{U},\leq)$ becomes a directed set. Now suppose, for a moment, that G is separable. Then G, being locally compact, is metrizable and hence there is a decreasing sequence (U_n) of open symmetric relatively compact sets in G which is a base for the neighbourhoods of e. So one may take $\mathcal{U} = \{U_n : n \in \mathbb{N}\}$, and then $(\mathcal{U},\leq)$ is order isomorphic to the positive integers and any net indexed on $(\mathcal{U},\leq)$ can be regarded as a sequence. Returning to the general case, for each U in $\mathcal{U}$ let ϕ_U be that positive multiple of the function $r \mapsto \Delta(r)^{-\frac{1}{2}} 1_U(r)$ satisfying $\|\phi_U\|_1 = 1$. Thus $(\phi_U)_{U \in \mathcal{U}}$ is a net of self-adjoint elements in the unit ball of $L^1(G)$. Let U_0 be some fixed member of $\mathcal{U}$, and consider a function f in $L^1(G)$. Given an $\varepsilon > 0$, there will be a continuous function ϕ on G which has compact support and satisfies $\|f-\phi\|_1 < \varepsilon$. Let C be the support of ϕ and a the measure of the relatively compact set $U_0 C$. As ϕ is uniformly continuous there will be a set U in $\mathcal{U}$ with $U \subset U_0$ and such that $|\phi(s^{-1}r) - \phi(r)| < \frac{\varepsilon}{a}$ for all $s \in U$ and all $r \in G$, and so

$$\|\phi_V * f - f\|_1 \leq \|\phi_V * (f - \phi)\|_1 + \|\phi_V * \phi - \phi\|_1 + \|\phi - f\|_1$$
$$\leq 2\|f - \phi\|_1 + \int |(\phi_V * \phi)(r) - \phi(r)|dr$$
$$< 2\varepsilon + \int |\int \phi_V(s)[\phi(s^{-1}r) - \phi(r)]ds|dr$$
$$\leq 2\varepsilon + \iint \phi_V(s)|\phi(s^{-1}r) - \phi(r)|drds$$
$$< 3\varepsilon$$

whenever $V \in U$ and $V \geq U$. This shows that $(\phi_U)_{U \in U}$ is an approximate identity for $L^1(G)$. □

A *representation* of G on a Hilbert space H is by definition a strongly continuous homomorphism of G into the group of unitary operators on H. As the weak and strong operator topologies on $L(H)$ induce the same topology on the group of unitary operators on H (see Lemma 22.3), a weakly continuous homomorphism of G into the group of unitary operators on H is actually a representation of G on H. The *trivial representation* of G on H is the function carrying each element of G onto the identity operator on H. The *left* and *right regular representations* of G are the representations λ and λ' of G on $L^2(G)$ defined by the formulae

$$[\lambda(r)f](s) = f(r^{-1}s) \quad \text{and} \quad [\lambda'(r)f](s) = \Delta(r)^{\frac{1}{2}}f(sr)$$

respectively, where r and s in G and f in $L^2(G)$ are arbitrary (in verifying that λ and λ' are weakly continuous one uses the fact that the continuous complex-valued functions with compact support on G are dense in $L^2(G)$). One defines *direct sums, subrepresentations, intertwining operators, quasi-equivalence, equivalence,* and *disjointness* of representations of G and one defines a representation of G to be *irreducible, multiplicity-free*, a *factor representation* or *factorial*, and *of type* x, $x = I, II_1, II_\infty, III$, by merely replacing R by G in the corresponding definitions in Appendix B.

There is a well-known and important construction which allows one to deduce a large part of the representation theory of locally compact groups from that of involutive Banach algebras. Briefly put, this construction shows that there is a natural one-one correspondence between the representations of a locally compact group G and the nondegenerate representations of the involutive Banach algebra $L^1(G)$ (cf. Theorem C.2). To describe the construction, consider a representation π of G on a Hilbert space H. It follows easily from the Cauchy-Schwartz inequality and the Riesz representation theorem that for each element f of $L^1(G)$ there is a unique continuous linear operator $\tilde{\pi}(f)$ on H satisfying

$$\langle \tilde{\pi}(f)x,y \rangle = \int f(r)\langle \pi(r)x,y \rangle dr \tag{1}$$

for all $x,y \in H$, and one can show that $\tilde{\pi}$ is actually a nondegenerate representation of $L^1(G)$ on H. Indeed, $\tilde{\pi}$ is obviously a linear map from $L^1(G)$ to $L(H)$, and is even a representation of $L^1(G)$ on H as

$$\langle \tilde{\pi}(f)^*x,y \rangle = \int \bar{f}(r)\langle x,\pi(r)y \rangle dr$$

$$= \int \Delta(r^{-1})\bar{f}(r^{-1})\langle \pi(r)x,y \rangle dr$$

$$= \int f^*(r)\langle \pi(r)x,y \rangle dr$$

$$= \langle \tilde{\pi}(f^*)x,y \rangle$$

and

$$\langle \tilde{\pi}(f)\tilde{\pi}(g)x,y \rangle = \int f(r)\langle \pi(r)\tilde{\pi}(g)x,y \rangle dr$$

$$= \iint f(r)g(s)\langle \pi(r)\pi(s)x,y \rangle ds\, dr$$

$$= \iint f(r)g(r^{-1}s)\langle \pi(s)x,y \rangle dr\, ds$$

$$= \int (f*g)(s)\langle \pi(s)x,y \rangle ds$$

$$= \langle \tilde{\pi}(f*g)x,y \rangle$$

for all $x,y \in H$ and all $f,g \in L^1(G)$. To show that $\tilde{\pi}$ is nondegenerate, take a vector x in H with $\tilde{\pi}(f)x = 0$ for all $f \in L^1(G)$. Then $\langle \pi(\cdot)x,x \rangle$, being a continuous bounded function satisfying

$$0 = \langle \tilde{\pi}(f)x,x \rangle = \int f(r)\langle \pi(r)x,x \rangle dr$$

for all $f \in L^1(G)$, must be identically zero, and so $0 = \langle \pi(e)x,x \rangle = \|x\|^2$ in particular. Finally, notice that one can recover π from $\tilde{\pi}$ by means of the formula

$$\langle \pi(r)\tilde{\pi}(f)x,y \rangle = \langle \tilde{\pi}(\lambda(r)f)x,y \rangle \tag{2}$$

where $r \in G$, $f \in L^1(G)$, and $x,y \in H$ are arbitrary and where, for all such r and f, $\lambda(r)f$ is defined by $(\lambda(r)f)(s) = f(r^{-1}s)$, $s \in G$.

Conversely, given a nondegenerate representation ρ of $L^1(G)$ on H, there is a representation π of G on H satisfying

$$\pi(r)\rho(f)x = \rho(\lambda(r)f)x \tag{3}$$

for all $r \in G$, $f \in L^1(G)$, and $x \in H$. Indeed, from the nondegeneracy of ρ and the fact that

REPRESENTATIONS OF LOCALLY COMPACT GROUPS

$$\|\rho(\lambda(r)f)x\|^2 = \langle\rho((\lambda(r)f)^**(\lambda(r)f))x,x\rangle$$
$$= \langle\rho(f^**f)x,x\rangle$$
$$= \|\rho(f)x\|$$

for all $r \in G$, $f \in L^1(G)$ and $x \in H$ it is easy to see that there is a homomorphism π of G into the group of unitary operators H satisfying (3). The strong continuity of π follows from Proposition B.1 and the fact that $r \mapsto \lambda(r)f$ is, for each $f \in L^1(G)$, a continuous function from G to $L^1(G)$ (to prove this one again uses the fact that the continuous functions with compact support on G are dense in $L^1(G)$). One can now form the representation $\tilde{\pi}$ of $L^1(G)$, and can even prove that $\tilde{\pi} = \rho$ as follows. Given vectors x and y in H, there will (by Proposition B.1) be a function θ in $L^\infty(G)$ with $\int f(r)\theta(r)dr = \langle\rho(f)x,y\rangle$, $f \in L^1(G)$. Thus for any three functions f, g, and h in $L^1(G)$ one has

$$\langle\tilde{\pi}(f)\rho(g)x,\rho(h)y\rangle = \int f(r)\langle\pi(r)\rho(g)x,\rho(h)y\rangle dr$$
$$= \int f(r)\langle\rho(h^**(\lambda(r)g))x,y\rangle dr$$
$$= \iint f(r)\theta(s)(h^**(\lambda(r)g))(s)dsdr$$
$$= \iiint f(r)\theta(s)h^*(t)g(r^{-1}t^{-1}s)dtdsdr$$
$$= \iint \theta(s)h^*(t)(f*g)(t^{-1}s)dtds$$
$$= \langle\rho(h^**(f*g))x,y\rangle$$
$$= \langle\rho(f)\rho(g)x,\rho(h)y\rangle$$

using Fubini's theorem, and therefore $\tilde{\pi} = \rho$ by the nondegeneracy of ρ.

THEOREM C.2. Let G be a locally compact group. For each Hilbert space H the formulae (1) and (2) establish a one-one correspondence $\pi \longleftrightarrow \tilde{\pi}$ between the representations of G on H and the nondegenerate representations of $L^1(G)$ on H. If H and K are two Hilbert spaces and if π and ρ are representations of G on H and K, respectively, then

(a) an operator from H to K intertwines π and ρ if and only if it intertwines $\tilde{\pi}$ and $\tilde{\rho}$,
(b) $\pi(G)'' = \tilde{\pi}(L^1(G))''$,
(c) π and ρ are quasi-equivalent if and only if $\tilde{\pi}$ and $\tilde{\rho}$ are, and
(d) $(\pi_E)^\sim = (\tilde{\pi})_E$ for all projections E in $\pi(G)'$.

Proof. The one-one correspondence has already been established, and assertions (a)-(d) follow without difficulty from (1), (2), and continuity of *-isomorphisms of von Neumann algebras with respect to the ultraweak topology (cf. Proposition A.3(iii)). □

It is easy to verify that if π is the trivial representation of G on a Hilbert space H then $\tilde{\pi}(f) = (\int f(r)dr)I_H$, $f \in L^1(G)$, and that if λ is the left regular representation of G then $\tilde{\lambda}(f)g = f*g$, $f \in L^1(G)$ and $g \in L^2(G)$.

By modifying the proof of the last theorem a little one can obtain the following technically-useful result.

PROPOSITION C.3. A Borel homomorphism of a locally compact group into the group of unitary operators on a separable Hilbert space is a representation.

Proof. Let π be a Borel homomorphism of a locally compact group G into the group of unitary operators on a separable Hilbert space H. Then (1) still serves to define a representation $\tilde{\pi}$ of $L^1(G)$ on H, and π and $\tilde{\pi}$ satisfy (2). If x is a vector in H with $\tilde{\pi}(f)x = 0$ for all f in $L^1(G)$ then (1) implies that for each fixed vector y in H one has $<\pi(\cdot)x,y> = 0$ a.e. with respect to Haar measure on G. But then there must be an r in G with $<\pi(r)x,y> = 0$ for all vectors y in some countable dense subset of H, and so x = 0 and $\tilde{\pi}$ must be nondegenerate. Then by Theorem C.2 there will be a representation ρ of G on H with $\tilde{\rho} = \tilde{\pi}$, and thus (by (2)) with $\pi = \rho$. □

REFERENCES

1. W. Arveson, *An Invitation to C*-Algebras*, Springer-Verlag, New York, 1976.

2. L. Auslander and C. C. Moore, *Unitary representations of solvable Lie groups*, Memoirs Amer. Math. Soc., No. 62, Providence, 1966.

3. N. Bourbaki, *Topologie générale*, chap. IX, Act. Sc. Ind., No. 1045, Hermann, Paris, 1958.

4. D. Bures, *Abelian subalgebras of von Neumann algebras*, Memoirs Amer. Math. Soc., No. 110, Providence, 1971.

5. J. Dixmier, Sur la réduction des anneaux d'opérateurs, Ann. Ec. Norm. Sup., 68 (1951), 185-202.

6. J. Dixmier, Sur les C*-algèbres, Bull. Soc. Math. Fr., 88 (1960), 95-112.

7. J. Dixmier, Sur les structures boréliennes du spectre d'une C*-algèbre, Publ. I.H.E.S., 6 (1960), 297-303.

8. J. Dixmier, Dual et quasi-dual d'une algèbre de Banach involutive, Trans. Amer. Math. Soc., 104 (1962), 278-283.

9. J. Dixmier, *Les C*-algèbres et leurs représentations*, Gauthier-Villars, Paris, 1964.

10. J. Dixmier, *Les algèbres d'opérateurs dans l'espace Hilbertien (Algebres de von Neumann)*, deuxième édition, Gauthier-Villars, Paris, 1969.

11. N. Dunford and J. T. Schwartz, *Linear Operators, Part I: General Theory*, Interscience Publishers, New York, 1964.

12. E. G. Effros, A decomposition theory for representations of C*-algebras, Trans. Amer. Math. Soc., 107 (1963) 83-106.

13. E. G. Effros, Convergence of closed sets in a topological space, Proc. Amer. Math. Soc., 16 (1965), 929-931.

14. E. G. Effros, The Borel space of von Neumann algebras on a separable Hilbert space, Pacific J. Math., 15 (1965), 1153-1164.

15. E. G. Effros, Transformation groups and C*-algebras, Ann. Math., 81 (1965), 38-55.

16. E. G. Effros, Global structure in von Neumann algebras, Trans. Amer. Math. Soc., 121 (1966), 434-454.

17. E. G. Effros, The canonical measures for a separable C*-algebra, Amer. J. Math., 92 (1970), 56-60.

18. J. A. Ernest, A decomposition theory for unitary representations of locally compact groups, Trans. Amer. Math. Soc., 104 (1962), 252-277.

19. J. Feldman, Borel sets of states and of representations, Michigan Math. J., 12 (1965), 363-366.

20. J. Glimm, Type I C*-algebras, Ann. Math., 73 (1961), 572-612.

21. R. Godement, Sur la théorie des représentations unitaires, Ann. Math., 53 (1951), 68-124.

22. A. Guichardet, Une charactérisation des algèbres de von Neumann discrète, Bull. Soc. Math. France, 89 (1961), 77-101.

23. A. Guichardet, Caractères des algèbres de Banach involutives, Ann. Inst. Fourier, 13 (1963), 1-81.

24. A. Guichardet, Sur un problème posé par G. W. Mackey, C. R. Acad. Sc. Paris, 250 (1966), 962-963.

25. P. R. Halmos, *Measure Theory*, van Nostrand, Princeton, 1950.

26. F. Hausdorff, *Grundzüge der Mengenlehre*, Chelsea, New York, 1943.

27. H. Hoffmann-Jørgensen, *The theory of analytic spaces*, Various Publication Series, No. 10, Aarhus Universitet Matematisk Institut, Aarhus, 1970.

28. R. V. Kadison, Normalcy in operator algebras, Duke Math. J., 29 (1962), 459-464.

29. R. R. Kallman, A generalization of free action, Duke Math. J., 36 (1969), 781-789.

30. R. R. Kallman, Groups of inner automorphisms of von Neumann algebras, J. Functional Anal., 7 (1971), 43-60.

31. C. Kuratowski, Sur le théorie des fonctions dans les espaces métriques, Fund. Math., 17 (1931), 273-282.

32. C. Kuratowski, *Topologie I*, deuxième édition, Monografie Matematyczne, Tom 20, Warsaw, 1948.

33. C. Kuratowski, *Topologie II*, Monografie Matematyczne, Tom 21, Warsaw, 1950.

REFERENCES

34. C. Lance, Refinement of direct integral decompositions, Bull. London Math. Soc., 8 (1976), 49-56.

35. L. H. Loomis, *An Introduction to Abstract Harmonic Analysis*, van Nostrand, Princeton, 1953.

36. N. Lusin, Sur la classification de M. Baire, C. R. Acad. Sc. Paris, 164 (1917), 91-94.

37. N. Lusin, Sur les ensembles analytiques, Fund. Math., 10 (1927), 1-95.

38. G. W. Mackey, Induced representations of groups, Amer. J. Math., 73 (1951), 576-592.

39. G. W. Mackey, Induced representations of locally compact groups. II. The Frobenius reciprocity theorem, Ann. Math., 58 (1953), 193-221.

40. G. W. Mackey, *The theory of group representations*, mimeographed notes, University of Chicago, 1955; reprinted as *The Theory of Unitary Group Representations*, University of Chicago, Chicago, 1976.

41. G. W. Mackey, Borel structure in groups and their duals, Trans. Amer. Math. Soc., 85 (1957), 134-165.

42. O. Maréchal, Topologie et structure borélienne sur l'ensemble des algèbres de von Neumann, C. R. Acad. Sc. Paris, 276 (1973), 847-850.

43. F. I. Mautner, Unitary representations of locally compact groups. I, Ann. Math., 51 (1950), 1-25.

44. F. I. Mautner, Unitary representations of locally compact groups. II, Ann. Math., 52 (1950), 528-556.

45. F. I. Mautner, Note on the Fourier inversion formula on groups, Trans. Amer. Math. Soc., 78 (1955), 371-384.

46. M. A. Naimark, Factor-representations of a locally compact group, Soviet Math. Dokl., 1 (1960), 1064-1066.

47. M. A. Naimark, *Normed Rings*, revised American edition, P. Noordhoff N. V., Groningen, The Netherlands, 1964.

48. J. von Neumann, Einige sätze über messbare abbildungen, Ann. Math., 33 (1932), 574-586.

49. J. von Neumann, On rings of operators. Reduction theory, Ann. Math., 50 (1949), 401-485.

50. O. A. Nielsen, The asymptotic ratio set and direct integral decompositions of a von Neumann algebra, Canadian J. Math., 23 (1971), 598-607.

51. O. A. Nielsen, Borel sets of von Neumann algebras, Amer. J. Math., 95 (1973), 145-164.

52. G. K. Pedersen, *C*-Algebras and Their Automorphism Groups*, Academic Press, London, 1979.

53. J. Schwartz, Type II factors in a central decomposition, Comm. Pure and Appl. Math., 16 (1963), 247-252.

54. J. T. Schwartz, *W*-Algebras*, Gordon and Breach, New York, 1967.

55. M. Souslin, Sur une définition des ensembles measurables B sans nombres transfinis, C. R. Acad. Sc. Paris, 164 (1917), 88-91.

56. M. Takesaki, *Tomita's theory of modular Hilbert algebras with its applications*, Lecture Notes in Mathematics, No. 128, Springer-Verlag, New York, 1970.

57. V. S. Varadarajan, Groups of automorphisms of Borel spaces, Trans. Amer. Math. Soc., 109 (1963), 191-220.

INDEX OF NOTATION

As is customary, $\mathbb{N}$ denotes the positive integers, $\mathbb{Q}$ the rational numbers, $\mathbb{R}$ the real numbers, and $\mathbb{C}$ the complex numbers. Unless otherwise stated, the domain of a sequence is $\mathbb{N}$. If H is a Hilbert space then I_H (or just I if H is clear from the context) will denote the identity operator on H, $C(H)$ the scalar multiples of I_H, and $L(H)$ the set of all bounded linear operators on H. Here is a list of some of the more specialized notation together with the pages on which they are defined:

$f(\mu)$, $v(\mu)$, $A(\mu)$ 9,16,18

$L^2(\mu;H)$, $L^2(\mu;H)$ 14

ℓ^2, ℓ_n^2 22

$L^2(\mu;H,\alpha)$, $L^2(\mu;H,\alpha)$ 33

$\int_Z^\alpha H(\zeta)d\mu(\zeta)$, $\int_Z^\oplus H(\zeta)d\mu(\zeta)$ 23,24

$\int_Z^\alpha v(\zeta)d\mu(\zeta)$, $\int_Z^\oplus v(\zeta)d\mu(\zeta)$ 23,24

$\int_Z^\alpha A(\zeta)d\mu(\zeta)$, $\int_Z^\oplus A(\zeta)d\mu(\zeta)$ 23,24

$\int_Z^\alpha \pi(\zeta)d\mu(\zeta)$, $\int_Z^\oplus \pi(\zeta)d\mu(\zeta)$ 45

$\int_Z^\alpha A(\zeta)d\mu(\zeta)$, $\int_Z^\oplus A(\zeta)d\mu(\zeta)$ 75

$vN(H)$ 67

P_k, $P_{1,n}$ 85

E_k, $E_{1,n}$ 85

$vN_k(H)$, $vN_{1,n}(H)$ 86

Rep R, $\text{Rep}_n R$, Irr R, $\text{Irr}_n R$, Fac R, $\text{Fac}_n R$ 115

$\hat{R}$, $\widehat{R}$ 115

INDEX

A

Analytic subset   2
Approximate identity
   definition of   148
   sequential   149

B

Borel field
   of Hilbert spaces   22
   of representations   45
   of von Neumann algebras   74
Borel function   1
Borel generating sequence   74
Borel isomorphism   1
Borel measure   2
Borel space
   analytic   2
   definition of   1
   product   2
   quotient   2
   standard   2
   sum   2
Borel structure
   analytic   2
   countable generated   2
   countable separated   2
   definition of   1
   generated by   1
   product   1
   quotient   2
   relative   1
   standard   2
   sum   1

Borel subspace   1

C

C*-algebra   145
Canonical Borel measure   130
Central decomposition
   of a representation   46
   of a von Neumann algebra   76
Coanalytic subset   2
Coherence
   constant   22
   definition of   22
   equivalence of   28
Compatible   63
Constant field of Hilbert spaces   22
Convolution   151
Cross-section   2
Cyclic vector   143

D

Decomposable operator   18, 23
Diagonalizable operator   18, 23
Direct integral
   of Hilbert spaces   26
   of representations   45
   of von Neumann algebras   76
Direct integral decomposition
   of a Hilbert space   32
   of a representation   45
   of a von Neumann algebra   76
Dixmier's transversal theorem   8
Dual   176

E

Effros Borel structure 67
Ergodic Borel measure 118

F

Faithful 142
Field
   of Hilbert spaces 22
   of representations 45
   of von Neumann algebras 74

G

Glimm's theorem 115

H

Hausdorff
   Borel structure 66
   metric 64
   topology 65

I

Involution 145
Involutive Banach algebra
   definition of 145
   of type I 115

K

Kaplansky density theorem 140
KMS-condition 143

M

Mackey Borel structures 115
Maximal decomposition 46
Measurable set 2
Modular automorphism group 143

O

One-parameter group of automorphisms
   definition of 143
   strong continuity of 143
Operator field
   Borel 23
   definition of 22

P

Polish space 2
Positive linear functional 142
Projection
   abelian 140
   central support of 140
   equivalence of 140
   finite 140
   infinite 140

Q

Quasi-dual 115

R

Representation
   definition of 145, 153
   direct sum 146, 153
   disjoint 146, 153
   equivalent 146, 153
   factor, factorial 146, 153
   irreducible 146, 153
   left regular 153
   multiplicity-free 146, 153
   nondegenerate 146
   quasi-equivalent 146, 153
   right regular 153
   subrepresentation 146, 153
   trivial 153
   type of 146, 153
   zero 146

S

Schur's lemma 148
Separating vector 143
Separation theorem 5
Smooth dual 115
State 142

T

Totally bounded 63
Trace 142
Transversal 2

U - Z

Universally measurable 2
Vector field
   Borel 23
   definition of 22

Von Neumann algebra
   automorphism   143
   continuous   140
   definition of   139
   finite   141
   infinite   141
   inner automorphism   143
   outer automorphism   143
   properly infinite   141
   purely infinite   141
   reduction of   141
   semi-finite   141
   spatial isomorphism   140
   type of   141
Von Neumann density theorem   139
Von Neumann-Mackey cross-section theorem   10

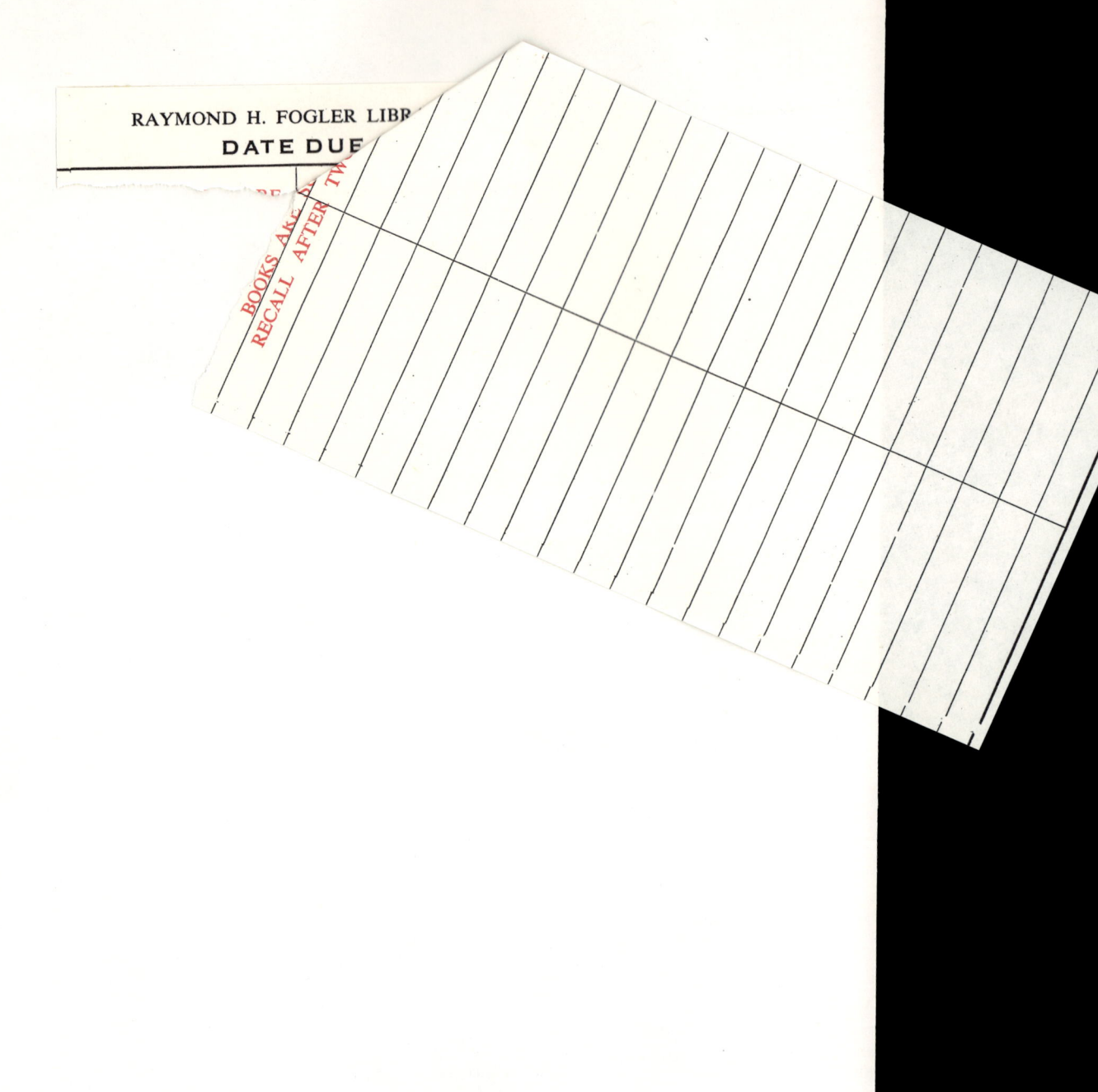